Vehicle Fine Fini

Vehicle Fine Finishing

PETER CHILD
MIMI

BSP PROFESSIONAL BOOKS
OXFORD LONDON EDINBURGH
BOSTON PALO ALTO MELBOURNE

First published 1987

British Library
Cataloguing in Publication Data
Child, Peter
Vehicle fine finishing.
1. Motor vehicles—Painting
I. Title
629.2'6 TL154

ISBN 0-632-01872-0

BSP Professional Books
Editorial offices:
Osney Mead, Oxford OX2 0EL
(*Orders:* Tel. 0865 240201)
8 John Street, London WC1N 2ES
23 Ainslie Place, Edinburgh EH3 6AJ
52 Beacon Street, Boston
Massachusetts 02108, USA
667 Lytton Avenue, Palo Alto
California 94301, USA
107 Barry Street, Carlton
Victoria 3053, Australia

Set by V & M Graphics Ltd.,
Aylesbury, Bucks
Printed and bound in Great Britain by
Billing & Sons Ltd., Worcester

Contents

Preface

In today's automotive market the need for and expectation of high quality levels is now greater than it has ever been in the past. The demand for excellence has also grown in the after sales and service market, and one item that has a high priority is paint durability and finish.

New and exciting developments in paint technology have brought great improvements to the surface coating of mass produced cars, and if they require refinishing then new skills and application techniques must be employed to restore vehicles to their original condition.

Vehicle Fine Finishing is for all those professional painters who use skill and expertise in refinishing motor vehicles, and I hope that they may find encouragement and useful information to enable them to carry on the rewarding task of painting motor cars.

Acknowledgements

I gratefully acknowledge all the help and kind consideration shown to me by the following, without whose help and co-operation this book could not have been written:

A. M. Stevens Esq
Aston Martin Lagonda Ltd
Ault Wiborg Paints Ltd
Copperleaf Cars Ltd
D. Kitchen Esq
DeVilbiss Co. Ltd
Glasurit Valentine
Arthur Holden & Sons Ltd
ICI Paints Division
J.R.Taylor & H.Foster Esq
Mebon Paints Ltd
Parsons Paints Ltd
R. Lynham Esq

1

Background

In the early days of motor vehicle construction the coachbuilder and painter realised that their expertise and talent could be transferred to the new, noisy enfant, with considerable gain to themselves. With the exciting emergence of the vehicle it was obvious that high standards of finish would help to sell the motoring concept to the wealthy who could indulge in the new craze.

The coachbuilders' skills came to the fore to construct a carriage that did not fall to pieces whilst being driven at speed, and afforded some protection to the occupants.

Open vehicles were the norm and only when the professions, doctors in particular, began to use the vehicle for business, did the closed coupé become a standard construction.

Coach painting methods were transferred directly to the vehicle bodies and the early vehicles were finished in the same, rather long-winded way as coaches, with the resultant hold-up in production procedures.

As coach painting disciplines were used directly, the early vehicles were often heavily lined; even spoked wheels had intricate lining and finishing pieces, laboriously applied.

At this time, all paint was applied by brush, and the type of paint used was the traditional heavy pigmented synthetic finish. This was overlaid with varnish at the end of a long and laborious finishing process, which gave high film builds and an acceptable finish for the day.

Application of materials by spray brought a new dimension into the finishing of vehicles, with the ability to coat up with an even build and to give a flowed out, high gloss gun finish and of course a vastly increased speed of application.

The motor industry really took off when Henry Ford decided to put the world on wheels. What a success story! Ford development then and now has contributed so much to the automotive industry and to the ordinary people of the world. The whole thing has been an exciting, revolutionary and highly emotional experience for mankind.

It would be an eye-opener indeed if every person who owned a motor car had to design, build and then paint one. The full extent of Ford's achievements, and those of so many other great motor men who followed, would be fully recognised.

The one fact that they all accepted was that every vehicle had to be painted. The better the paint finish and trim, the easier the car was to sell. The 'look' of the car was recognised to be of paramount importance. This is where pride in possessions comes to the fore – and the ladies are not shy in coming forward when colour, trim and style are to be selected.

Mechanicals are men's things; just ask any lady. If in doubt, stand in any car showroom and listen to the customer's wife say, 'I like the colour', and even if the car is too long for the garage or too expensive to insure, the sale will be completed. The marketing men will confirm this any time. Finish and appearance will always satisfy the customer's needs.

It was at this stage, therefore, that vehicle painting was dragged forward into the limelight – where it has remained ever since.

2

Early Systems

At first, car manufacturers sold the customer the mechanical piece, which in the case of Rolls-Royce included the radiator. The rolling chassis was then delivered to the coachbuilder of choice, Barclays or Mulliners.

Designs were offered to the customer, he made his decision, and the coachbuilders began. Luxurious trim and final paint finish enhanced the vehicle standard for the owner who then presented the machine to the many admirers. The rivalry between the coachbuilders did a great deal to raise the standards of their art, and in the final colour finish, perfection was the aim.

Even as low volume production began, it was obvious that the coachbuilders could not cope with demands and soon car manufacturers had commenced the construction of their own bodies. The top of the range cars, however, still retained the coachbuilders and it was only after the Second World War that Rolls-Royce and Bentley began to manufacture their own bodies.

Normally car bodies were constructed in wood, usually ash, and were covered in aluminium or mild steel sheet, rolled to contour and then fitted to the wood frame. Often leatherette or fabric was fitted over the panel work. This helped to hide imperfections in the panels and gave a uniformity to the car.

When vehicles were fully painted, it was a long and laborious job. Both aluminium and steel would be roughly abraded before a brush coating of zinc chromate. Heavy coats of hard chalk-based filler would be applied, followed by hand flatting and then the operation would be repeated until a surface suitable for undercoat was achieved. Undercoating took place, followed by further wet flatting. Colour was eventually applied by brush and left to dry through. As the materials were synthetic they dried by oxidation, which caused long delays in the paintshop. The final finish was flatted back to receive coats of copal varnish to enhance and protect the colour film.

In 1928 Sir Herbert Austin had the idea of finishing a vehicle in nitrocellulose. Sir Frederick Crane Ltd supplied a black nitrocellulose to Austin called Cranco. An Austin 7 was spray-finished in this material and overnight there was a revolution in production-line finishing.

The spray finishing of vehicles has developed from teams of paint sprayers operating sprayguns on either side of the track, to fully computerised automatic sprayguns set up in rows of traversing lines as vehicle bodies pass through, up to the present day when robotics have taken a further step forward.

Devilbiss Co. Ltd have developed a seven-stage robotic spray painting set-up. The vehicle body is brought into a spray painting bay on rails and is located against buffers. This calls for accurate positioning of the body on the carrying trolley as the robots are programmed to spray accurately to very tight parameters. The robots all spray together and the vehicle is coated inside and out within seconds. The paint finish is totally even and every part of the shell is painted.

Every aspect of the building of motor vehicles is being computerised and the construction is by robots, but the after-sales finishing and repairs must still be done by a highly skilled person. The art of fine finishing belongs to a skilled man, who takes pride in his work and his ability to follow the paint manufacturer's instructions - and add that little extra skill and technique during application which give the added dimension to the job.

This has been the case from the very first days of vehicle finishing. However, there are, and have been, painters and painters. Like chefs, they can interpret and project their skill, and customers are willing to pay to have the expertise lavished on their car. This has always been the stage at which emotion and the love of the car enters the picture. A perfectly finished Aston Martin or Bentley will look stunning, appealing and exotic; poorly painted it will look drab, spoilt and a lot less desirable.

On production-line vehicles before the Second World War finishes were generally the alkyd melamines that were developed for high stoving and they continued in use up until the development of acrylics in the '50s. The pre-war finishes were heavy-bodied pigmented materials that gave opacity, high build and a heavy appearance, made worse by the manufacturer's need to produce vehicles at a high rate, and hence the rather 'dry' application by the sprayers which gave the orange peel effect. Runs and sags were to be avoided at all costs as these vehicles had to be removed to the 'hospital line' for flatting and refinishing.

Although nostalgia tells everyone that cars were painted to higher standards in the past, the fact is that modern paint technology and application methods give today's customer a far superior product than he ever enjoyed before. One only has to examine basecoat and clear systems that have been applied to Jaguar cars to realise the achievements in painting mass-production motor vehicles.

3

Standards and Tests of Today

The technology behind all the major car manufacturers today is quite outstanding, with the emphasis on vehicle finishing obvious to all observers.

The introduction of electrophoretic primers in mass production ensured even coating in primer by electrical deposition over every part of a monocoque shell, even to between spot welding. The build of between 1 and 1½ thou encapsulates the steel structure and gives the maximum protection against rust, as well as a firm substrate for subsequent filler and colour coats.

The development of acrylic lacquers from the canning industry in the USA during the '50s gave the motor industry a lacquer that was impervious to ultraviolet light and the bombardment that takes place on any paint film during a life-time exposed to the elements. Colour pigment was built into acrylic lacquer, which is a water white material. Test panels, put on Florida exposure as well as accelerated weathering and saline spray, proved to be virtually unmoved by the harsh treatment.

Nowadays every major manufacturer is using acrylic finish in high stoving (320°F) in two forms, either in thermosetting finish (one-bake system) or thermoplastic (bake-sand-bake).

The finish obtained on Ford cars is excellent and nibs and dirt inclusions are non-existent. The basecoat and clear systems offer remarkable finishes that are both extremely attractive and very hard-wearing indeed. Repair of these paint films is very straightforward.

GM Vauxhall as well as Jaguar have favoured TPA (bake-sand-bake) which enables the shell to be coated and then, after stoving, to be flatted back with fine paper and hydrocarbon oil, which removes any dirt inclusion, runs or sags. The vehicle is then rebaked, which causes a wetting up of the film to a very high gloss, totally free of imperfections. This finish is superb and will in normal conditions last for many years. However, small or part repairs must be carried out with care as there does exist a degree of surface tension in the paint film as a result of reflow. It is most important not to load up the surface with too much solvent as crazing can occur.

At the very top of the market only two manufacturers stand well ahead of all the competition: the legendary Rolls-Royce and Aston Martin. Both companies offer the greatest skills and expertise to their customers in the standard of finish of their vehicles. Both companies use TPA low bake and both spend many hours of labour in achieving outstanding coachwork and finish. (At the 1984 Motor Show Aston Martin Lagonda were awarded two golds for coachwork, one for the Lagonda and the other for the Volante.)

The finishes on these cars need care and it is always hoped that customers will give the car the care and attention that is needed to keep it in the pristine condition in which it leaves the showroom.

Of the leaders, Aston Martin Lagonda have the disadvantage of working the cosmetic panels in aluminium, with all the attendant problems of welding and fixing to a mild steel frame and floorpan. Aluminium is soft and requires careful attention to shape and rollering into body contours. The sheer size of panels on the Lagonda give rise to many problems in construction and painting but these are all overcome, to the great credit of all those involved in the manufacture of the finest and fastest production sports saloons in the world. A vehicle fit only for James Bond and connoisseurs of excellence! Figure 1 shows the perfect gloss and panel work of a Lagonda.

Fig. 1 High gloss finish and panel work of a Lagonda

Fig. 2 Aston Martin Volante Convertible in white aluminium awaiting final abrading before going to the paintshop for cleaning and priming with self-etch primer. Note the inner structure has already been prepared and coated with styrene/acrylic black underbody coating

Rolls-Royce, with their reliability and long life, offer a different light on motoring. The cars are opulent, beautifully trimmed and finished out. The paint process is a little easier due to the use of steel panels in the construction of the car. There is less movement in the panel and generally stretching and shrinkage are not a problem. Fine finish on steel is therefore much easier.

Both manufacturers clean the vehicle body down thoroughly and then proceed with self-etch primer followed by two-pack fillers. Flatting back takes place, with any small filling up imperfections such as file or body hammer blemishes. A sealer coat is applied, and then first colour. Aston Martin then put the shell on the line for mechanical build and trim. When this operation is completed the vehicle returns for second paint. This consists of a complete flat back, masking up, and final coating in TPA with clear lacquer. The car then begins a road test programme and when everything is tuned up to perfection the car is flatted back, and stone chips or road blemishes are sprayed in, then the car is polished to a brilliance throughout. After final PDI the car is then ready for delivery to the customer.

Tests

Often painters are not fully aware of the tests that paint manufacturers set up to obtain the highest degree of performance from their product. The vehicle manufacturers are instrumental in ensuring that paint is tested to the very limit before giving approval for materials to be supplied on the line.

As an example, the Florida exposure test proves a paint system's durability in the harshest environment. Panels are painted in the full schedule for original equipment and are exposed in Florida, where there exists high humidity, a salt-laden atmosphere and an ultraviolet light bombardment situation that is severe in the extreme. Panels are inspected and at the point of failure are returned to the paint makers. UK and European motor manufacturers demand a timed exposure in Florida of up to five years.

The stability and durability of the paint film are of prime importance and original finishes are designed to meet the challenge of the day-to-day use of a motor vehicle in any climatic condition.

Severe cold can cause problems to a paint film and the cold crack test takes a painted panel down to −30°C and then tests the elasticity of the film.

These extremes of heat and cold give the paint manufacturers the very worst case conditions for measuring the performance of their product. Again, the long-term durability and effect can only be achieved using the correct procedures and materials.

The question is often asked of how long a paint film should last. In a European climate, fifteen or twenty years of good service could be expected, provided the owner of the vehicle takes care of the film. Washing the car down regularly and often will ensure that salt, road film and grit remain on the surface for a minimum of time, and garaging a vehicle adds to the life-span of the finish.

Care and good maintenance pay dividends all the time, and ensure that the owner has a car to be proud of, the refinisher has a vehicle that is easier to repair, i.e. good, uncontaminated surface to work on in the event of accident repair, easy colour identification and matching, and, when the vehicle is sold, a better price allowance for the owner.

Colour

Contrary to popular belief it is the stylists at the motor manufacturer's who choose the colours, not the paint companies. When the decisions have been made for the colour range for a particular year the paint manufacturers who normally supply colour to the line are informed and sent a 6″ × 4″ panel, known as a 'master', with the colour applied over a primer substrate. From this master, a sample batch is made up in the high stoving acrylic and presented to the motor manufacturer for test and approval. The samples are supplied both in wet and dry form, and are subjected to rigorous tests, of opacity (the hiding power), gloss, weatherability, adhesion, and impact resistance. If all goes well then the paint manufacturer is notified that the colour can be accepted on the line and in due course orders will follow. In manufacture the paint will be tested several times, and very thoroughly before filling out. Samples of that batch are retained by the paint manufacturer, whilst other samples are sent to the motor manufacturer prior to acceptance on the line. When the final approval is given, the material is then supplied and used. The paint is checked at so many stages by both the manufacturer and the user that it ensures a 100% material is sprayed onto the vehicle and the owner of the vehicle is assured of the quality of the finish of his car.

The time factor

So often today, standards demanded in a repair paint shop are reasonable in final finish but unreasonable as far as time goes. Customers who damage their vehicles require the car back yesterday. Unfortunately they are able to communicate this desperate need either to the reception engineer or to the manager, who of course believe (a) that the customer is always right, (b) that he is the king because he makes paydays possible, and (c) that a service industry has to behave in that way.

Well, I understand the position but do not always necessarily agree. The world and his wife do not understand, and never have understood, the refinishing of a motor vehicle. The paintshop is some dirty hole round the back of the garage full of 'Picassos' and black dogs who live in a perpetual world of gloom and filth. Added to which, in everybody's eyes they only coat up; all the work is done by the bodyshop and mechanics. I have reminded many garages and refinishers that paintwork and bodywork are like a saddlemaker and blacksmith – very separate trades, but it is all on the horse!

When car owners have been made to understand the process that has to be carried out to make their pride and joy look whole again, they step back in amazement with comments such as, 'I never knew!', 'Good heavens – what a lot of work!', 'Really?', and 'Wait till I tell them at the golf club!'.

In a simple sentence the customer needs to be educated; but it really does not stop there. From works supervisors to middle management to directors, they all need educating – some more than others. I suppose poor communication has been the torment of vehicle refinishing ever since it began. Management, so often trained as engineers or bodymen, have consistently failed to understand the paint procedure and the length of time necessary to achieve a high standard. They fail to remember that car owners generally only see the outside finish and the interior trim of their vehicle; what the back axle looks like remains forever a mystery.

Paintwork that is hurried and carelessly applied due to outside pressure returns in due course for 'warranty rectification'. What a total waste of time and money! You do the same job twice or even three times because the owner wanted it on Friday. He is not the expert in the building or repair of his vehicle – he is the end user. He must be clearly advised that repairs take time and he must be patient. It is far better to tell someone at the beginning that the job will take five days, and be able to accomplish the work in that time, than promise three days and struggle and rush to complete a second-rate job. Many times painters have said to me, 'just one more day and I could have made a lovely job of that one'. What a sad but true comment. There is a deep lesson to be learnt there, in both customer relations and reputation building.

Great expectations

So, what should a customer expect from his repaired motor car? Where do we set the standards, parameters and criteria? It seems evident that no one really knows. 'Good commercial finish' is the usual cry. Does this mean, it is OK and will last for a year? Or, it will last as long as the original finish on the vehicle? Has a customer a right to come back in eighteen months complaining about a fault in the finish?

I suggest he has a right, but he may be unreasonable. So often I have been told to 'tidy' a car up, as the owner is '... going to sell it; he had a little knock recently and his wife is nervous travelling in it, you understand, keep it down to a price ...' etc. Two years later the car returns – 'What a lousy job, rust showing through already' – 'But it was only a tidy up'. – 'Ah, yes, well, his wife liked it so much when she saw it she decided to keep it!'. There is another whole book to be written covering the stories that all painters have truly experienced.

The only way to avoid these misunderstandings is to give guidelines of straightforward information, and to educate the owner as to what he will get for his money. If he pays he can have a Rolls-Royce finish on the wing of his Skoda – but it will not appear the same as the rest of the car! People get colour matching and appearance matching confused: the cry 'I can see it has been repaired' tells you this.

If you take any of the major manufacturers as an example you will find that main agents will refinish an accident damaged vehicle as closely as possible to the original finish. They have invested huge sums in equipment and facilities to ensure that they can do so.

Training of personnel is another area where I am glad to see great improvements have been made. The more skilled people that we have in paintshops, the better the performance from the industry.

The standards are rising and hopefrully this trend will continue. Owners will be delighted with their vehicles, management in paint and body shops will be pleased with turn-round, and executives will smile at the profits.

However, so often a manager will insist that only what is on the job card must be done. Well, for mechanics and body men this is fine, and it clarifies the position for them, but (and it is a big but) painters are in a quite unique position here because the all-important paint finish must be acceptable to the owner. A blemish or, more important, a mismatch of colour, and in metallic poor appearance or texture, will have a customer refusing to accept his vehicle. Often with metallic finishes it is just not possible to conclude the repair as an edge to edge. It is important to take a wider view, and that can certainly save a great deal of time and money.

This was brought home to me vividly when as a paintshop foreman I was presented with a very expensive two-door sports car with the passenger door skin quite badly damaged. The customer had been quoted for the repair of the door only and the job card called for painting the panel only. The problem was the colour – it was a special orange metallic!

After attempting to match the colour using the mixing scheme the final tinting was done by eye and spraying out 6 inch × 4 inch panels and putting these into the low bake oven to develop and cure. It was not possible to match the colour with the tinters available although the appearance and texture of the metallic were fairly accurate. I then decided to paint the whole side of the vehicle up to the roof line. After careful preparation the car was painted once and finished out. The customer was delighted with the job and paid his bill. The plus points far outweighed the minus because the vehicle was masked, prepared, painted and low baked once – it went straight through the system. There were no hold-ups, except for colour matching which is an infill job within the shop, and we had a satisfied customer. Our reception engineer had allowed an extra hour for colour matching and so although the job took about an hour longer to mask up and wet flat the other panels, and approximately an extra litre of colour was used, the time saved and profit on the job were considerable.

Sensible application of the skills within the paintshop will enable quality work to be accomplished repeatedly and with few problems. It must be understood that the owner of a vehicle sees only the exterior paintwork and the interior trim.

The first impressions are those that the owner remembers always. If these impressions are good, very little in the way of criticism will follow. However, the reverse is that if he finds one thing wrong then he will go over the vehicle with a fine toothcomb. I have witnessed this many times first-hand.

Very occasionally a wealthy customer will arrive with a newly acquired but dilapidated vehicle and say, 'Please rebuild it to perfection and I don't mind how much I spend'. Everyone rubs their hands – the proprietor for the profit, the manager for the use of a job where staff can be employed in a slack moment, and the paint and body shop where at last they can take their time and get it right.

It has been my experience that a customer will pay if he gets the job he wants. The important point here is to ensure that the job is properly programmed. An old heap in the corner covered with dust tends to attract hours booked on it without a lot happening and the vehicle is there so long that it becomes a nuisance rather than a job to be done.

Make sure that everyone, including the customer, knows fairly precisely what is going to happen. Then let it be seen to happen. For instance, all the brightwork to be removed and sent out for rechroming; seats to be removed and sent to the trimmers; as these items return, everyone is motivated into playing their part without too much pressure. Everyone takes a pride in a job well down, and should be fully involved and enjoy the customer's praise at the end.

I was involved in the painting of a 3.8 E-type roadster, which was trimmed in cream leather and finished in Jaguar British Racing Green after a total mechanical and body rebuild. When the car was ready for collection by the customer everyone involved came to look at the car. It raised morale by several points, and when the customer arrived he was ecstatic. It was a very successful and profitable operation that had been well planned, and delivered on time.

The team

The importance of planning, team co-operation, involvement and utilisation of the skills within the shop cannot be over-emphasised. Well trained and experienced staff should very well paid and very well stretched. Management must never fall into the trap of believing that staff can be easily replaced. It can take six months or more to find out if a replacement person is suitable and his work consistent and of a high standard. So often good people have moved on because of poor salaries and the replacement has cost the company dear. It is far better to build an integrated team with high salaries and hold them with structured increases. Balances have to be struck, however, and staff must come to terms with the reality of the commercial world, which in modern times consists of the whims of the clearing bank manager and the company accountant (both of whom, in my experience, lost touch with reality when they left school). Reputations, as well as high incomes, have to be earned. Both can be easily lost and the skills of management should be to be economically effective as well as ensuring that the business of repairing and refinishing vehicles to high standards is well maintained and proceeds into the future.

4

Equipment

Fine finishing can be achieved in several ways, but most important of all is the equipment that will be used. The DeVilbiss Company has been the world leader in the development and supply of all spraying and finishing equipment. The JGA gun is used throughout the world for the application of all types of paint.

At present the developments in automotive spraying robotics have revolutionised production spraying, and overall their continuing efforts have made a significant contribution to the finishing and refinishing industry.

To paint any vehicle it is essential to have, in perfect working order:

- A spray booth or low bake oven
- A compressor of adequate capacity
- A spraygun
- An air regulator
- A viscosity cup and time clock

The range of ovens and spray booths is such that any refinisher should find a unit that will suit his needs. The manufacturers are pleased to advise as to layout and installation and will commission the booth as well as carrying out the balancing of the air movement. Figure 3 shows a typical spray booth and oven.

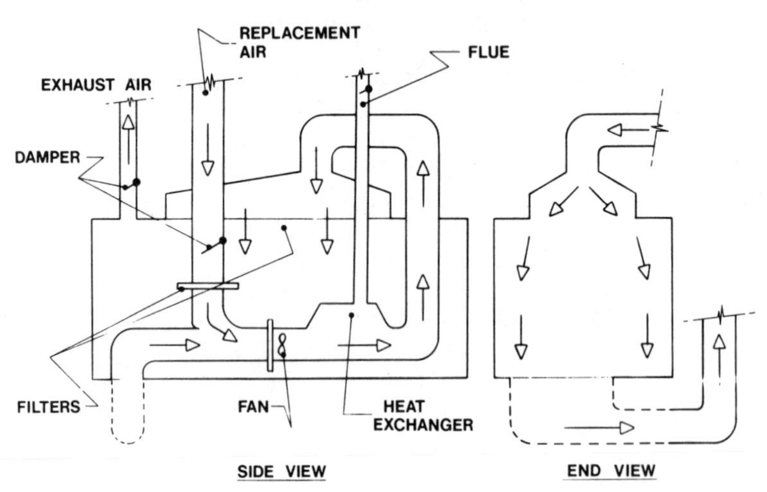

Fig. 3 A low bake oven

The spraygun

Let us consider the spraygun first. This is the tool used for spraying and is made in various sizes, classified according to its capacity to atomise and apply volumes of paint. Sprayguns are classed as high production, medium production and low production instruments and consequently the selection of the gun depends on how much work it is expected to carry out, i.e. how fast it will apply the paint. The type of spraygun that is chosen for refinishing therefore depends largely on the amount and type of work to be done.

Generally speaking, however, the standard refinishing spraygun is the type JGA (suction or pressure feed) (Fig. 4) and the other guns used are the type GFG (gravity feed) or, for smaller areas or smaller output, the type TGA (suction feed), the type MPS (gravity feed), and the type CGA (suction feed).

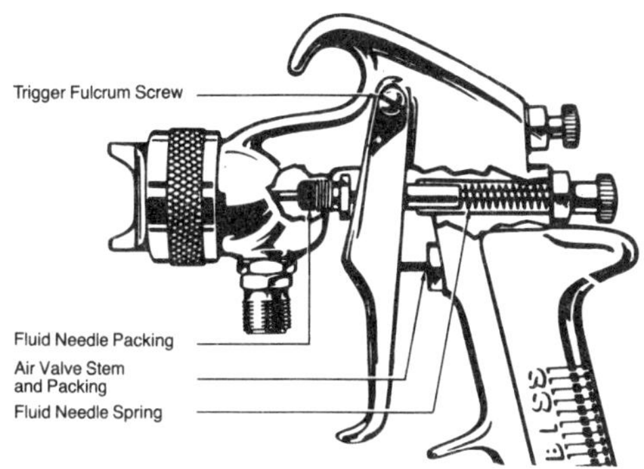

Fig. 4 A DeVilbiss JGA spraygun

Sprayguns are of the separate container or attached container types and these two types can be further divided into suction, gravity or pressure feed, bleeder and non-bleeder, external and internal mix guns.

Suction feed gun

This is a type of spraygun in which the stream of compressed air siphons material from the container attached to the spray head of the gun. This gun is usually limited to quart size containers or smaller and is easily identified by the fluid tip extending slightly beyond the air cap. Suction feed guns are used where there are many colour changes and they are the most popular in the car refinishing industry.

Gravity feed spraygun

This spraygun is fed by force of gravity from a paint cup attached to the top of the body. It is particularly suitable for quick colour changes and easy filling. The type GFG gravity feed spraygun uses the same air cap and fluid tip combinations as the type JGA suction feed spraygun and achieves as fine an atomisation as a pressure feed gun.

Pressure feed gun

This is a spraygun to which the paint is forced by pressure from a tank, fluid cup or pump. The gun is fitted with an air cap and fluid tip combination that does not siphon the paint. The fluid tip is generally flush with the air cap. A pressure feed gun is used when large amounts of the same material are being sprayed, the material is too heavy to be siphoned from a container by suction, or when extra-fast application is required. It is sometimes used in the car refinishing trade when large amounts of work need to be done very quickly, particularly on commercial vehicles. Pressure feed is used by nearly all car manufacturers.

One great advantage to be gained by using pressure feed from a separate container is that the spraygun is not handicapped – it can be moved about much more freely, it can spray upwards, downwards, even upside down if you like, without having to worry about spilling paint from a cup. A recently developed 2-quart remote cup is really a pressure feed tank in miniature which provides the refinisher with the means of doing a car more quickly and achieving a smoother, glossier finish than with an attached paint feed cup.

Spraygun set-ups

The phrase 'set-up' is a very important one in spray painting and includes three items, which can be called the principal parts of the spraygun (see Fig.4.). They are the parts that are changed to make the spraygun more adaptable for applying different kinds of material. They are changed according to whether suction or pressure feed is being used and to match the output volume of the air compressor.

Do not conclude from this that a spraygun set-up has to be continually revised and corrected. The above items which require changes are brought to your attention to show you what factors have to be considered in providing universal set-up combinations. These combinations are all listed in the DeVilbiss publication, *Spray Gun Selection Guide.*

The air cap

What is the air cap? The air cap is the part on the extreme working end of the gun. It is attached to the body of the gun by a threaded ring. There are two types of air cap:

- External mix. This has a number of air outlet holes in it, and the air and paint mix beyond the air cap, causing atomisation. This is the more commonly used air cap.
- Internal mix. This air cap has a slot in it and the air and paint mix behind the slot thus atomising the paint inside the air cap. Used with low pressure, low air consumption guns.

Air caps are specified by a number which is stamped on the front of the cap.

The spraygun, of course, uses air. The air is passed through the air cap and, after doing its job, escapes into the atmosphere. The amount of air used (or consumed) by an air cap is called the air consumption of the cap and is measurable in terms of cubic feet per minute (c.f.m.). The amount of air used by the air cap varies with the air pressure. For example, a No. 30 cap will consume 7.4 c.f.m. at 30 lb pressure, 9 c.f.m. at 40 lb and 10.6 c.f.m. at 50 lb. Therefore the higher the spraying pressure the greater the volume of air required.

At this point we must not lose sight of the relationship between the air compressor and the spraygun. It is a vital relationship because the air compressor is the supplier of the volume of air (c.f.m.) that the air cap consumes. In other words, the air compressor in a spray painting set-up has to be large enough in capacity to satisfy the demands of the air cap.

The DeVilbiss *Spray Gun Selection Guide* lists the air required at different pressures for air caps. This guide is therefore also a basis for air compressor selection.

The fluid tip

The fluid tip is the part of the spraygun directly behind the air cap. It is the nozzle through which the paint is directed into the airstreams coming out of the air cap. The fluid tip is a seat for the fluid needle and these two act as a valve controlling the flow of paint. Fluid tips are made in many sizes and the size reference is to the centre hole of the tip. Fluid tip sizes are: E, EX and D for suction feed, and FF, FX and FZ for pressure feed. To determine the size of fluid tip to be used for varying applications or materials, you should refer again to the DeVilbiss *Spray Gun Selection Guide.*

The fluid needle

The fluid needle is a long, pointed valve that seats into the fluid tip. Under the force of a spring it is normally pushed tight into the fluid tip opening but when the trigger is pulled the needle is withdrawn from the tip, thus allowing the paint to flow.

Fluid needles are made in as many sizes as fluid tips to coincide with them and the needle must always be of the same size as the fluid tip, i.e. an EX fluid tip requires an EX needle. The remainder of the spraygun's working parts are:

- The trigger, the action of which not only releases air into the air cap, but also withdraws the needle from the fluid tip allowing the paint to flow.
- The fluid adjustment screw, which controls the amount of travel of the fluid needle to let more or less paint through the fluid tip.
- The air valve, which is opened by the action of the trigger.
- The spreader adjustment valve, which controls the air to the holes in the horns of the air cap and regulates the size of the spray pattern from maximum width down to a narrow or round pattern.

Air and fluid hose

Two kinds of hose are used in the average spray painting system, one to convey air to the spraygun and the other (in pressure feed) to take paint to the spraygun. Whilst air and fluid hose are not made of the same materials they are both constructed of a synthetic rubber inner tube overlaid with a braided cover and finally covered with rubber. This is then vulcanised into a unit which is strong and yet flexible. Fluid hose has to be made of the right synthetic rubber

Minimum pipe size recommendations

	Air pressure drop at spray gun (lb)					
Size of air hose Inside diameter	5-foot length	10-foot length	15-foot length	20-foot length	25-foot length	50-foot length
$\frac{1}{4}$ inch						
at 40 lb pressure	6	8	$9\frac{1}{2}$	11	$12\frac{3}{4}$	24
at 50 lb pressure	$7\frac{1}{4}$	10	12	14	16	28
at 60 lb pressure	9	$12\frac{1}{4}$	$14\frac{1}{4}$	$16\frac{3}{4}$	19	31
at 70 lb pressure	$10\frac{3}{4}$	$14\frac{1}{2}$	17	$19\frac{1}{2}$	$22\frac{1}{2}$	34
at 80 lb pressure	$12\frac{1}{4}$	$16\frac{1}{2}$	$19\frac{1}{2}$	$22\frac{1}{2}$	$25\frac{1}{2}$	37
at 90 lb pressure	14	$18\frac{3}{4}$	22	$25\frac{1}{4}$	29	$39\frac{1}{2}$
$\frac{5}{16}$ inch						
at 40 lb pressure	$2\frac{1}{4}$	$2\frac{3}{4}$	$3\frac{1}{4}$	$3\frac{1}{2}$	4	$8\frac{1}{2}$
at 50 lb pressure	3	$3\frac{1}{2}$	4	$4\frac{1}{2}$	5	10
at 60 lb pressure	$3\frac{3}{4}$	$4\frac{1}{2}$	5	$5\frac{1}{2}$	6	$11\frac{1}{2}$
at 70 lb pressure	$4\frac{1}{2}$	$5\frac{1}{4}$	6	$6\frac{3}{4}$	$7\frac{1}{4}$	13
at 80 lb pressure	$5\frac{1}{2}$	$6\frac{1}{4}$	7	8	$8\frac{3}{4}$	$14\frac{1}{2}$
at 90 lb pressure	$6\frac{1}{2}$	$7\frac{1}{2}$	$8\frac{1}{2}$	$9\frac{1}{2}$	$10\frac{1}{2}$	16

Compressing outfit		Main air line	
Size (hp)	Capacity (c.f.m)	Length (ft)	Size (inches)
$1\frac{1}{2}$ and 2	6 to 9	Over 50	$\frac{3}{4}$
3 and 5	12 to 20	Up to 200	$\frac{3}{4}$
		Over 200	1
5 to 10	20 to 40	Up to 100	$\frac{3}{4}$
		Over 100 to 200	1
		Over 200	$1\frac{1}{4}$
10 to 15	40 to 60	Up to 100	1
		Over 100 to 200	$1\frac{1}{4}$
		Over 200	$1\frac{1}{4}$

Fig. 5 (a) Table of air pressure drop (b) Table of minimum pipe size recommendations

compound to resist the swelling, erosive and general deteriorating action of thinners and solvent, whereas air hose is made to contain pressure and resist only such contaminants in the air as water and oil in small amounts. To simplify identification air hose is coloured red and fluid hose black.

With the average production guns $^5/_{16}$ inch internal diameter air hose is used and for the smaller, low consumption guns $^1/_4$ inch internal diameter air hose. The fluid hose normally used is $^3/_8$ inch internal diameter. In pressure feed systems for conveying air from the air compressor to a pressure feed tank containing the paint, $^7/_{16}$ inch air hose is used.

It is important to remember that the friction of compressed air passing through hose causes a certain amount of pressure drop; the table in Fig. 5(a) shows this clearly. For lengths of air hose longer than 25 feet the next larger diameter hose should be used to minimise this pressure drop. (See Fig.5(b) for minimum pipe size recommendations.)

Hose connections

Hose connections are of the screw-on type and have a threaded nut which swivels on the connection body. This nut has threads matching the threads of the spraygun, and is screwed on and tightened with a wrench.

Quick-detachable DeVilbiss air and fluid connections are also available which enable instantaneous changes to be made. These connections incorporate an automatic valve which seals the hose effectively from air or paint seepage when they are disconnected.

The air transformer

The next item in a spray painting installation after the spraygun and hose is the air transformer, which is placed between the gun and the air compressor (see Fig.6). The air transformer can be described as a device which removes oil, dirt and moisture from compressed air, filters and regulates that air, indicates by gauges the regulated air pressure and provides outlets for sprayguns, air dusters and other air tools.

Air coming out of a compressor is what we could call raw. It is raw because it is generally at a higher pressure than that needed for spray painting, and it therefore has to be controlled down. It is raw because it is dirty. It will contain water vapour which condenses to water liquid, and it can pick up rust scale from any pipeline which is beig used. It can also have in it oil vapour from the compressor and it is therefore obvious that it needs not only regulating but cleaning too. An air transformer does precisely this.

The principal parts of an air transformer are a condenser, a filtering device, an air regulator, a pressure gauge or gauges, outlet valves and a drain cock. The air transformer, by means of the screw-down valve, reduces the main line air pressure to the pressure that you require for the spraygun; it cleans the air by means of filters, centrifugal force, baffles or expansion chambers; and it incorporates a drain cock at the bottom of the transformer chamber to run off any water collected by these means.

Air transformers are built to pass certain maximum volumes of air and it is

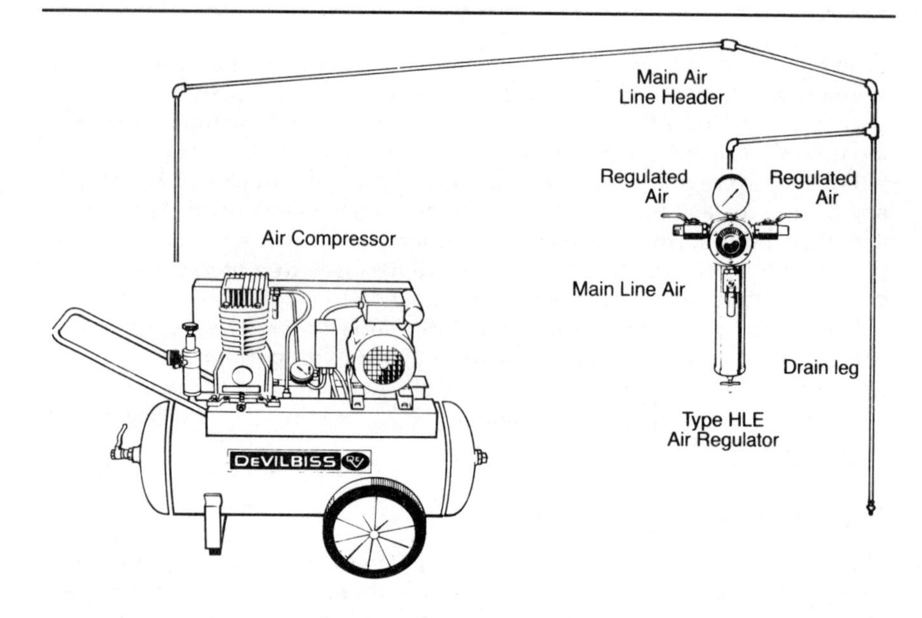

Fig. 6 A typical layout of air compressor, air line and air transformer

therefore necessary to ensure that the transformer fitted is of sufficient capacity to handle the volumes of air produced by the compressor being used. The maximum recommended capacity of type SJ transformer is 14 c.f.m. and that of type HLE is 50 c.f.m.

The air compressor

The two types of compressors commonly used in spray painting are the piston and the diaphragm type. Piston types are heavy duty and diaphragm types are for lighter service.

Compressed air costs money to produce and consequently air compressors require to have a high level of efficiency in order to run economically. Since an air compressor is designed to maintain a volume of air, its efficiency is called volumetric efficiency. The best compressors for painting are those with which the amount of air loss between the displacement and actual delivery is lowest. Displacement is the term applied to the theoretical volume of air which the compressor draws in, compresses and discharges. Free air delivery is the actual volume of air which is discharged by the compressor. The difference between these two is the air loss. Whether piston or diaphragm type, the air loss should not be more than 20–30 per cent and a good compressor should therefore be at least 70 per cent efficient.

Details and descriptions of air compressors can be found in the DeVilbiss *ABC of Spay Equipment*.

Paint heating

The advantages of spraying heated cellulose lacquers, cellulose synthetics, and synthetic enamels include: better finishes – the improved flow-out of heated material produces a smoother, glossier surface with no orange peel or sagging, and rubbing down is often unnecessary; more uniform coatings – atmospheric conditions cannot affect viscosity and film build; lower air pressure – the viscosity of paint falls as its temperature rises, so that a lower air pressure is required for atomisation, and overspray is minimised; faster application – one hot coat is often as thick as two cold coats and repeated spraying operations with intervening drying periods are thus avoided (Fig.7).

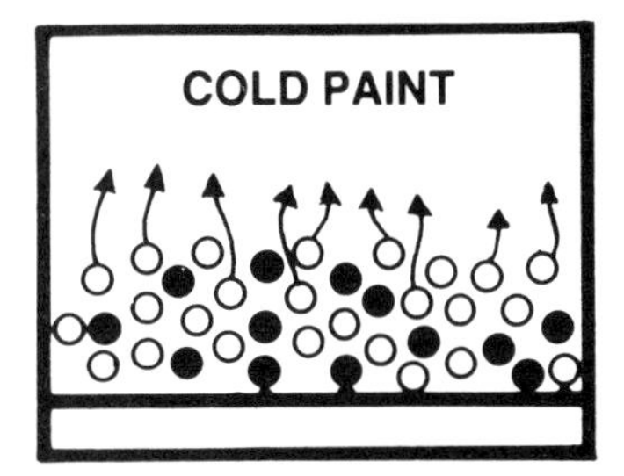

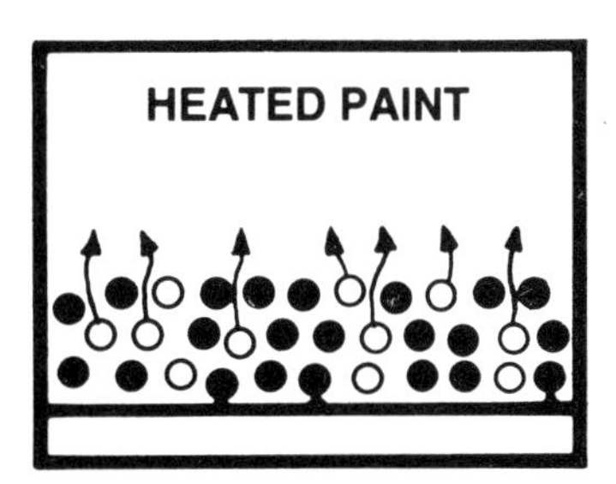

Fig. 7 Hot and cold paint

The reason why a thicker coat can be obtained is simple. When any paint has been evenly applied, it is desirable that it becomes more viscous, or less liquid, as quickly as possible so that it will not flow out too much on the surface or form runs, sags or fatty edges. Although some solvents are driven out of cold applied paints by the atomising action, the bulk of the solvent remains to leave the material 'runny'. In contrast to this, with hot material the thinners evaporate more rapidly and the paint cools, causing it to return to about the same consistency as it was in the can and to resist any tendency to run or sag, even though a heavier coat is applied.

Hot spray cup

This cup has a capacity of one-quart and can be attached directly to a standard spraygun. The paint is heated to a temperature of 165°F by a thermostatically controlled element connected by flex to a suitable electric point. The flex is unplugged from the cup whilst spraying.

Type MH paint heater

A circulation hot water system heats the paint supplied from a pressure feed tank to the required temperature in a few seconds through a heat exchanger and heated material flows to the gun through a heat jacketed fluid hose.

The refinishing shop

Let us imagine that we have just walked into a well-designed and efficiently run small refinishing shop, and see how things are carried out.

The first thing that we notice is that the shop is rectangular in shape, not too high, and has at one end what is called a floor mount exhaust system. This is in effect a steel box with an opening in the bottom facing towards the shop which literally scoops the air up by means of an exhaust fan installed inside and discharges it outside the building either through a louvre shutter or by means of ductwork to a higher level. Sometimes a wall mount working on exactly the same principle is used.

It is obvious that if air is taken out of a spray shop by means of an exhaust fan, provision must be made for air to come in, otherwise the fan will attempt to make a vacuum in the shop and may damage itself. Therefore in the wall immediately opposite the exhaust outlet is an air inlet fitted with glass fibre filters, which clean the air as it passes through. The area of these filters is at least three times the overall area of the exhaust fan and they allow air to pass through freely, replacing that taken out by the fan.

The floor of the shop will be clean and well swept, without a litter of scraps of rubbing down or masking paper, old paint tins and the like. The walls will be smooth finished in white paint to reflect light and to stop the collection of any dust or overspray.

The air compressor will be mounted outside the room, preferably in its own brick-built compartment, and the air line which runs into the spray shop will be pitched at an angle so that any moisture that may collect in it will run back to the air compressor so as to be easily drained off from the drain cock at the bottom of the air receiver. The correct oil level is maintained in the compressor sump and the air intake filter is cleaned at least once a week. The air transformer will be bolted securely to the side of the spray shop as near to the operator as possible so that he can easily read the guages and operate the valves. It will be at least 25 ft from the air compressing plant and the take-off for it will be from the top of the air line, or from the upper part of a vertical drain leg dropping from the main air line. This is an additional insurance against water in any amount being passed down to the transformer. The drain cock at the bottom of the transformer must be opened at intervals to allow any accumulated moisture to be blown out.

Where space allows, the ideal spraying conditions are provided by a car refinishing booth which ensures that no dust or dirt in any form can reach the work being sprayed and is the perfect method of refinishing vehicles.

If the refinisher wishes to apply low bake enamels, and gain the advantages of higher gloss, greater durability and savings of time, he must also install an oven. Basically this consists of an enclosed insulated structure through which is passed a continuous current of hot air which 'stoves' the finish after about 40 minutes. A heating system is also utilised to raise the temperature of the adjacent spray booth to a satisfactory level for applying these materials. Whilst one car is being stoved another may be sprayed, allowing a continuous throughput of vehicles.

Colour mixing

The weight mixing scheme that has been developed to cope with the avalanche of new colours that were introduced by the motor manufacturers in the 1960s is one of the most significant developments in the vehicle refinishing industry.

To dispense colour by weight brought forward many problems. Firstly, the accuracy of the equipment itself. It had to have the ability to accurately weigh down to 5 grams over a weight range of 5 to 5,000 gms, which is approximately the weight of 5 litres of colour.

Secondly, the staining strength of the mixing enamels had to be constant for repeatability, and not suffer with batch wander on any area of standard. In the days of volume mixing the enamels were not always the same but it did not matter too much as the painter's skill was such that he tinted as necessary. Also, he had very few car colours, something like 800 were on offer in 1960. Today the figure is 18,000!

Mixing by weight is at least five times more accurate than by volume. Computers have now made the matching of new colours a matter of routine, and the coding of the weight formulation a straightforward affair.

Paint manufacturers offer a range of products in weight mixing form: nitrocellulose, which is still widely used in the UK, acrylic finish which possibly is not so popular, the two-pack isocyanate finishes which are used extensively on the continent and are gaining in the UK market now, and finally synthetic finishes, such as ICI Permobel, which is a full bodied gloss from the gun finish ideal for passenger vehicle respray and all commercial vehicles.

Operating a weight mixing scheme saves time and money. The time saved revolves around the ability to match any colour, in any quantity, there at the point of use without having to refer outside the spray shop.

The money saved is in the stock holding. Before the weight mixing scheme, colours were bought daily from a supplier. Usually, the paintshop over-ordered, in case of faults or rework. This material went into stock, 10 tins became 20, 20 became 50 and so on. The stock holding of colour became very expensive in most paintshops. Most of it ended up being thrown away. With the mixing enamels necessary to operate the scheme only approximately 40 need to be held in stock, although this varies from paint manufacturer to paint manufacturer.

As stylists and designers develop more colours, the paint manufacturers can offer the refinishing trade the colours instantly by updating the microfiche formulae. Weight mixing is here to stay, and the back-up that these systems can give the painter is immeasurable.

For example, many colours vary on the production line due to colour batch wander, stoving times or different suppliers. As many as seven variants can occur for one colour and the weight mixing scheme offers formulations for each variant. This must mean more accurate refinishing and faster turnround.

The fact that the formulation is known to the painter also means that, should the decision be made to alter a colour, the knowlege of precisely what enamels are in the colour make-up allow him to tint with those enamels. The more mixing enamels that are present in a colour the 'dirtier' it becomes. It is therefore important to ensure that no new enamel is introduced to the formula when tinting.

If he is to succeed then the modern refinisher must set up his shop with the latest technology that is available. Equipment that is modern and performs consistently is the keystone to success. In simple terms, no matter how good a paint sprayer may be, he cannot compensate for a poor spraygun, a low volume compressor, a faulty air separator and no spray booth. The right equipment, used properly, is the foundation for excellent work standards and profitable business.

Robots

Not only is it essential to get an even coating of paint over a panel for maximum life and durability, which makes painting vehicle shells a very laborious task; it is also a dirty and repetitive operation. These are some reasons why robots have been developed for mass-produced vehicles. In some cases the robots are set up on the line and the vehicle passes through, being sprayed by traversing heads which are carefully programmed by computer to traverse very accurately and deposit the exact film weight desired by the manufacturer. Other robotic set-ups consist of the vehicle being brought into a bay where up to seven robots will coat up the shell in a precise manner. This modular robot system is working well in the USA and has the benefit of speed and minimal dirt inclusions. Any vehicle that has a problem does not go onto the pick-up line, so ensuring that the flow of vehicles to the build up track is 100% perfect.

Banks of these modular stations are set up and the production rate from them is considerable as it can take as little as 80 seconds to coat up a family saloon inside and out.

The finish that robotics are able to achieve is extremely good in terms of film build, evenness and finish overall. This will automatically give a protected and long life to the paint film in all adverse conditions.

Customers of all types of vehicles demand reliability and long life from the vehicle, and finish is one area that comes under the most careful scrutiny.

Use of electrostatic equipment

Background

More and more manufacturers are turning to this type of equipment coupled with the introduction of robots on production lines to obtain a more controlled and perfect finish from the application procedure. In metallic finishes the evenness in the lay of the aluminium particles is of great importance as this determines the appearance and to some degree the texture of the film. By using electrostatic guns this even build is guaranteed as the charged particles of material are attracted to the workpiece and a certain amount of wrap around occurs, ensuring a coating of colour on door edges, etc.

This equipment does take out the skill of the operator to some extent, but at the same time it does ensure the evenness of film build.

Advantages

The advantages of using electrostatic spray equipment are:

- Greater material savings achieved by external charging of the paint.
- It is suitable for use with a wide range of standard coatings.
- It is reliable, easy to maintain and service, and simple to use.
- Maximum operating efficiency ensures a first class finish on localised work or in high production.
- Reduced spray booth maintenance due to higher transfer efficiency.

Equipment

The equipment necessary to carry out electrostatic spraying consists of:
A power supply.
A Manual Electrostatic Spray Gun.
The DeVilbiss Company Ltd manufacture a power supply unit known as the BFA-800. This unit has the voltage adjustable in two ranges: 0.60 kV manual application or 0.90 kV automatic application. LED displays provide continous monitoring of kilovolt and microamp readings. The unit has modular circuitry for easy servicing, and built-in protection against live voltage variations. To operate with this power supply DeVilbiss Company Ltd market a spraygun known as the BFL-800 Manual Electrostatic Spray Gun. This unit is reliable and safe to use and designed for ease of cleaning and maintenance. It has a high fluid flow rate and both round and fan spray caps can be used. The charged probe may be touched to any earthed objects without danger of sparking. For operator safety, current will only flow to the gun when the trigger is operated.

The technology advance that this type of equipment offers will certainly ensure that is more widely used in the finishing and re-finishing industry. Fig.8 shows a complete manual liquid system.

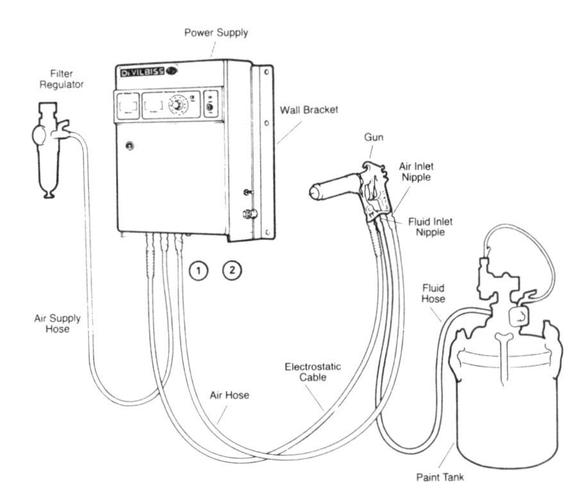

Fig. 8 Electrostatic equipment

Infrared

Infrared has played an ever-increasing role in the finish and refinish of the motor vehicle. All the major manufacturers are now using infrared curing in pre-colour areas, final finish or rectification line procedures.

The refinish industry has developed the use of infrared for part repairs and localised finishing, and it is rare to find it used in the full respray low bake conditions.

Most industrial finishes dry by a process of solvent removal or by polymerisation, both of which can be accelerated by the application of heat. Motor vehicle finishes having desirable properties such as hardness, flexibility, adhesion, chemical resistance and freedom from blemishes are easily obtained using electric infrared ovens. Stoving temperatures are rapidly attained and the results compare very favourably with those achieved by alternative heating methods, such as convection ovens.

The wide variety of electric infrared equipment now available is being exploited increasingly in the industrial finishing field for the following reasons:

- Rapid heat transfer to products shortens treatment times and increases throughput.
- Rapid heat up and cool down of heaters saves idling losses, and protects finished products during line stoppages.
- Highly suitable for mass production operations.
- Simple and accurate temperature control.
- No contamination from products of combustion.
- Equipment is relatively cheap and easy to install.
- Energy is used efficiently.
- Simply maintenance.
- Compact equipment saves factory space.

The radiant energy from an infrared source (the emitter) is directed towards the product to be heated, which ideally should have a large flat area and small thickness. However, the emitted energy can heat the whole surface of a three-dimensional product if it is suitably jigged. Uniform heating can be assisted by product rotation in a monorail oven and by multiple internal reflections of infrared energy within a highly reflective oven enclosure.

In the metals finishing field, conduction of heat throughout a complex product can reduce temperature gradients and thus play an important part in ensuring that shadowed corners or interior coated surfaces are properly heated.

Convected heat can make a secondary but important contribution to the heating process, depending upon the type of emitter used. This principle is used in the recently developed contraflow monorail oven which reverses the natural upward flow of hot air currents to give a downward flow over the products. This not only achieves better use of the convective components, it also tends to equalise the temperature of complex three-dimensional products.

Equipment

The energy emitted from the infrared source increases rapidly with its absolute temperature, obeying, in theory, a fourth power law. The wavelength of the radiant energy depends upon the temperature of the source, which, for industrial purposes, will be in the range of 600°C to 2200°C.

The selection of infrared equipment depends largely on the physical parameters of the process, but particularly on the nature of the material to be heated. Its colour, roughness, and chemical structure, collectively referred to as its 'absorptivity' or 'emissivity', have a marked effect on the wavelengths absorbed. It is therefore desirable to match the infrared heating system with the coating formulation to ensure the best results in terms of energy economy, productivity and surface quality.

Thermosetting paints require a minimum temperature to cure in a commercially convenient time and this can be generated by infrared radiation.

IR (infrared) emitters generate radiant energy in the 0.7–4 μm region and are characterised by their peak emission, short, medium and longwave. Medium and longwave are not as 'peaky' as short.

Organic and inorganic paint media have specific absorptions in the near IR due to overtones of principal absorption, molecular vibrations and charge transfer phenomena; these absorptions are dissipated as heat.

Cure of paint media may be isothermal (nominally) or by superheat raising the temperature until cure occurs. The great advantage of IR heating is the fast temperature rise, due to the large temperature difference between emitter and object. Superheat is a more effective way of using IR.

Factors affecting IR cure include: emitter type; emitter to product spacing; substrate thickness; substrate texture; paint colour and film thickness.

Dark colours heat faster than light colours. The rate of heating is related to the red absorbance of colours, e.g. white may attain 72°C in the time a dark blue takes to attain 87°C. A thicker white film heats up slightly faster than a thinner one, but this difference is not found with a black film. A rougher texture to the metal substrate may also cause a white film to heat up more quickly; again, a black film is not affected.

In isothermal IR cure, a saving in time to cure is confined to the heating-up time. In superheat cure, considerable savings in oven dwell time can be made, for example an alkyd MF paint stoving at 120° for 30 minutes in the isothermal mode can be cured in about five minutes in the superheat mode, depending on the kind of emitter and other factors.

Infrared cure is a valuable means of speeding up paint stoving, it is clean and the ovens and arches are very compact, but in installing units attention must be paid to the unique characteristics of radiation curing to obtain the best results.

5

Paint and Powder Coatings

In simple terms, paint is any fluid that can be applied to a surface which will dry to form a hard continuous film.

Three basic components make up paint:

(1) Pigment,
(2) Binder, and
(3) Solvent.

Added to these basics are the additives themselves.

Pigments (coloured pigment and extended pigments)

Pigments used in the manufacture of paint are very finely ground powders. These are either naturally occurring minerals or synthetic dyestuffs. They are fundamental to the system as they give the paint its hiding power (which is termed opacity) and its colour. They also contribute to the durability of the film and its length of life in the harsh environment of daily use in the world's climatic conditions.

In primers the pigments' main task is to resist corrosion and promote adhesion. In primer fillers they are selected to give good build and easy flatting. In final finish they are selected for a long lasting decorative effect.

Extenders are less expensive than pigments and are used to improve adhesion levels and to make flatting an easier operation.

Binder

This material is a film former and a polymer or reactive chemical plasticiser. It binds the particles of pigment together and provides good adhesion to the substrate. Adhesion is one of the most important factors in vehicle finishing because if the first coat of primer does not properly adhere to the steel, aluminium or GRP panel of a motor vehicle, all subsequent work is totally wasted.

Thinner (solvent diluent)

This makes the pigment and binder mixture fluid for use during manufacture and reduces it further for spray or brush application. Solvent evaporates from the paint film either by allowing it to air flash (21°C or 68° – 72°F) with air movement, or it is forced out by the application of heat (up to 80°C or 176° – 180°F) panel temperature.

Additives

These are small quantities of materials added to carry out particular functions, e.g. matting agents or accelerators.

Types of paints

Cellulose synthetic

These materials dry by solvent evaporation from the film at application. No chemical change takes place and the material dries out rapidly. The disadvantages are low build levels, high proportion of solvent (up to 60%) causing sinkage into the substrate, and the absence of chemical change means that the film becomes solvent when more coats are applied. Solvent retention can cause the film to remain soft for considerable periods.

Oil and synthetic based paints

In these materials initial drying takes place by solvent evaporation but final film hardening is due to a chemical change in the paint which is caused by the uptake of oxygen. Unfortunately this type of material suffers a critical recoat period during which wrinkling or lifting will occur if the film is repainted. This period commences after about three hours at normal paintshop temperatures (21°C - 68°F to 72°F) up to 36 hours.

What happens is that the paint film is gradually curing from the exterior surface down and is becoming less soluble in its own solvent.

Synthetic based paints have a high solid content (as much as 52%) and therefore their covering power and film build ability are very high.

Low bake

These materials will not air dry but need the application of heat to a panel temperature of 80°C (176°F) to effect a full cure and hardening of the material. There is an initial solvent flash off at application within the spray booth at 21°C (68°F - 72°F) and then a full cure at 80°C (176°F - 182°F) for up to 60 mins at that temperature, and then a cooling down with a certain uptake of oxygen which then makes the material non thermoplastic.

These materials are used for fast production levels in refinishing and they are the nearest thing to the vehicle manufacturer's original finish.

Two-pack materials

Paints of this group do not become a fully cured system until the paint and hardener are mixed together. The hardener is usually an isocyanate and is harmful, and therefore safety precautions must be observed when using the material. It is a very hard and durable film due to the catalyst action. It can be air dried or force dried and the vehicle can be very quickly put into use after application.

The growth in the use of these materials has been quite astounding and it appears that this type of paint is leading the way into the future.

Thermosetting acrylic enamel

This material is used extensively by the motor manufacturers as original equipment and it is important to fully understand some of the basics in terms of repainting vehicles surface-coated with this enamel. The properties of this material and the aspects of performance are clearly shown in the following text which has been taken in part from the paper presented to the Oil and Colour Chemists Association by Mr J.R.Taylor and Mr H.Foster.

The general chemistry of thermosetting acrylic resins is based on methylol acrylamide and hydroxylated acrylic monomers. Typical monomers which may be present in this type of material are given in a separate table with their contribution to the complete material. Such acrylic resins are suitable for use in many applications but have become well established in the automotive industry in the USA, Great Britain and Europe. These are specially suitable for the formulation of metallic finishes, since the gloss, polishing and weathering properties are usually of a very high order. It is usually possible to re-establish the original gloss of an exposed thermosetting acrylic enamel by burnishing with a mop and in this respect these finishes behave in a similar manner to nitrocellulose and thermoplastic acrylic enamels. The thermosetting acrylic resins may be thought of as a second generation which makes it possible to formulate paints having reflow properties similar to those based on thermoplastic resins.

Typical monomers which may be present in a thermosetting acrylic resin

Monomer	*Contribution*
Methyl methacrylate Styrene Vinyl toluene Acrylonitrile	Hardness
Ethyl Butyl 2-Ethyl hexyl Butyl maleate	Flexibility
Acrylamide Butoxymethyl-acrylamide Hydroxy alkyd acrylates Glycidyl acrylates Acrylic acid	Crosslinkages
Acrylic acid Methacrylic acid Maleic anhydride	Cure acceleration

Curing mechanisms of hydroxylated acrylic resins

The thermosetting acrylic resins now being used in the automotive finish industry are cured by the addition of a suitable melamine resin and stoved at temperature between 120 and 150°C for approximately 30 minutes, or at a lower temperature (80 to 90°C) after the addition of an acid catalyst such as phthalic acid or butyl hydrogen maleate. It is believed that for optimum properties, including excellent durability, the crosslinked film of such an acrylic melamine system is obtained when only a proportion of the total number of crosslinking sites have reacted. A study has been made of the rate of reaction of hydroxylated thermosetting acrylic resins stoved with butylated melamine resins, in order to ascertain the degree of crosslinking which occurs in reflow systems, (a) during the initial low bake (70 to 100°C), and (b) on final baking at normal stoving temperatures (120 to 150°C). The degree of cure was followed by the determination of the amount of material extracted from the cured film by methyl ethyl ketone in a Soxhlet apparatus.

Repair in process

Thermoplastic acrylic resins have been established for some time as suitable media in surface coating enamels in the automotive industry. One of their advantages has been the ability to reflow during the so-called repair in process techniques. Until recently it was assumed that thermosetting acrylic resins, because of their chemical nature, were not capable of being used in such processes, since during the initial stoving period flow would be prevented by excessive crosslinking of the reacting acrylic and melamine resins. It is, however, interesting to note that certain British, American and European car manufacturers have been interested in the possibility of using thermosetting acrylic resins in this process, in order to reduce the number of cars that have been returned to the finishing lines because of surface defects.

Two repair in process systems have been evaluated by car manufacturers: the bake sand bake, and the bake sand respray bake techniques. In both of these the enamel coat is sprayed and given an initial bake at a low temperature in the range of 70 to 100°C, at which stage solvent evaporation has taken place and with very little crosslinking so that the coating remains thermoplastic. The paint film must then be hard enough to permit defects to be sanded out with abrasive paper without excessive clogging. Subsequently the latter film is tak wiped and, in the bake sand bake process, immediately restoved at 140°C. During this final stoving the paint film must soften sufficiently to allow reflow to occur, thus obliterating the sanding marks. In the bake sand respray bake process the sanding areas are tak wiped and resprayed with a second coat of the enamel without any masking of the remainder of the car body. The paint is then given a final stoving at 140°C for 12 to 15 minutes. During the final stoving period all dry spray from the respray operation must be completely accepted by the first coat of enamel so that both coats are virtually indistinguishable.

The requirements of a reflow resin are summarised schematically in Fig.9,

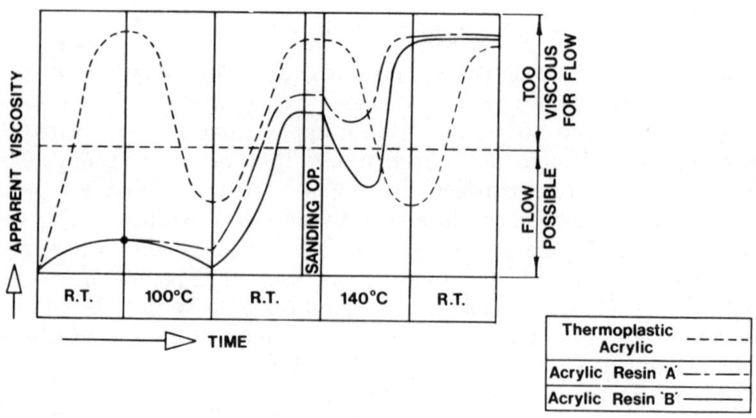

Fig. 9 Reflow of three types of acrylic resins

which illustrates the anticipated flow properties of three hypothetical resin systems:

- A typical thermoplastic acrylic resin (TPA).
- A typical thermosetting acrylic resin system (TSA).
- A reflow thermosetting hydroxylated acrylic resin system (RTSA).

Stage 1: Evaporation of solvent
During this stage the TPA resin, because of its high molecular weight, should reach its maximum hardness, whereas the TSA and the RTSA resins will still be comparatively soft at room temperature.

Stage 2: Low temperature bake
The TPA resin viscosity drops rapidly as the panel temperature increases. Both the TSA and the RTSA are thermoplastic at this stage but as the panel temperature increases, the crosslinking reaction commences, and the actual viscosity change is dependent upon both of these two properties.

RTSA resins are known to be slower curing at low temperatures than conventional TSA resins and therefore it is likely that the curves for these resins diverge as shown in Fig.9.

Stage 3: Removal from low temperature oven
As the panels cool down to room temperatures they become hard enough to sand. The TPA recovers its original solvent-free hardness (Fig.10).

Stage 4: Final high temperature bake
The initial stage of the final bake is critical. Both the TSA and the RTSA resins will flow to different extents, but in the case of the TSA resin the crosslinking reaction will take place too rapidly for reflow to occur, whereas the RTSA resin will have sufficient time to reflow before it becomes crosslinked.

Stage 5: Removal from high temperature
All three types of resin should attain a similar viscosity (hardness) after cooling to room temperature. It will be seen that the important reduction in viscosity

is that which occurs on the rebake after the sanding period, and attempts have been made to record viscosity changes in the region.

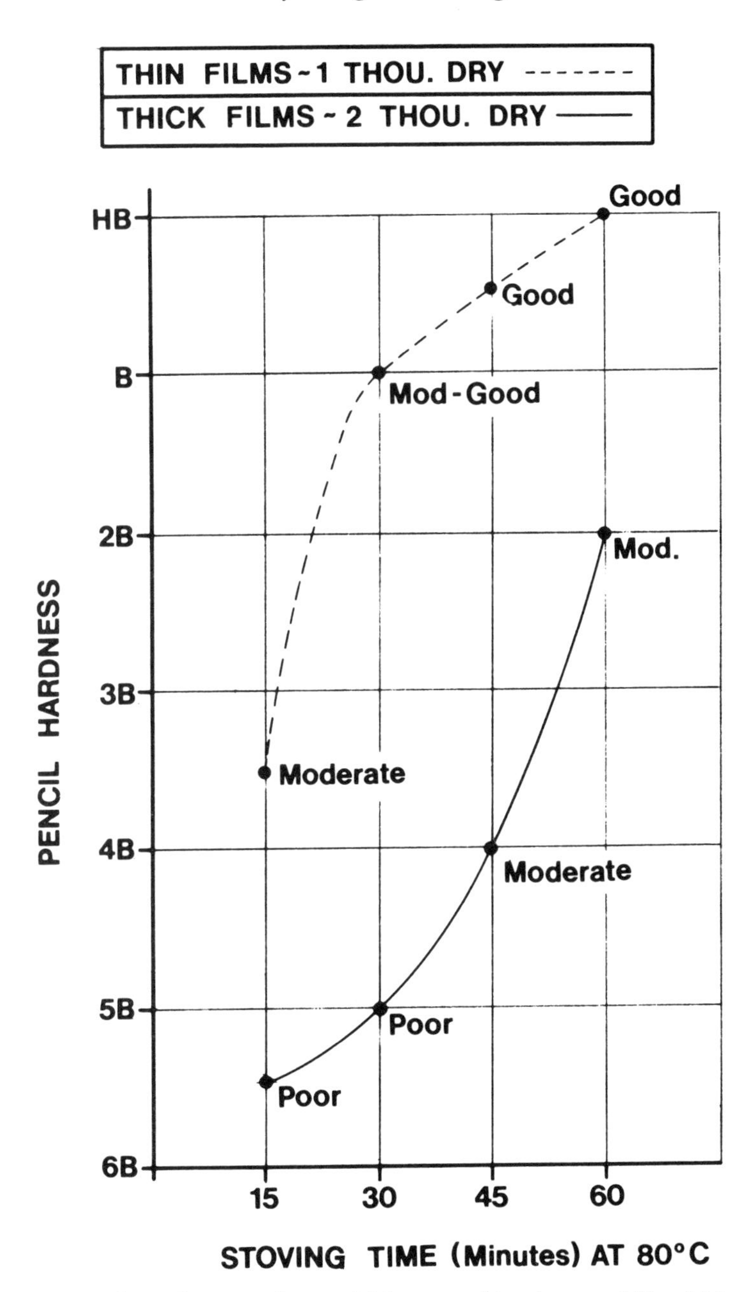

Fig. 10 Correlation between dry sandability, pencil hardness and film thickness of resin 'A'

Practical automotive specifications for reflow acrylic resins

The Ford specification ESB.M32J.102A was released in July 1967 and for the first time a standard for reflow performance was indicated to the resin and paint manufacturer. Initial low bake schedule was fixed at 15 minutes at temperatures ranging from 71 to 99°C (160 to 210°F) and in practice this meant that the enamel must be hard enough to sand after stoving for 15 minutes at 71°C but sufficiently thermoplastic to reflow and accept overspray after stoving for 15 minutes at 99°C. In addition the enamel was required to perform equally well after allowing 72 hours to elapse between initial bake and the final sanding operation.

Figure 9 shows the performance limits required of reflow thermosetting acrylic resins and the physical constants of two such resins, A and B. Type A represents a typical hydroxylated acrylic resin which when crosslinked with a melamine resin will only reflow after a comparatively low initial bake, and type

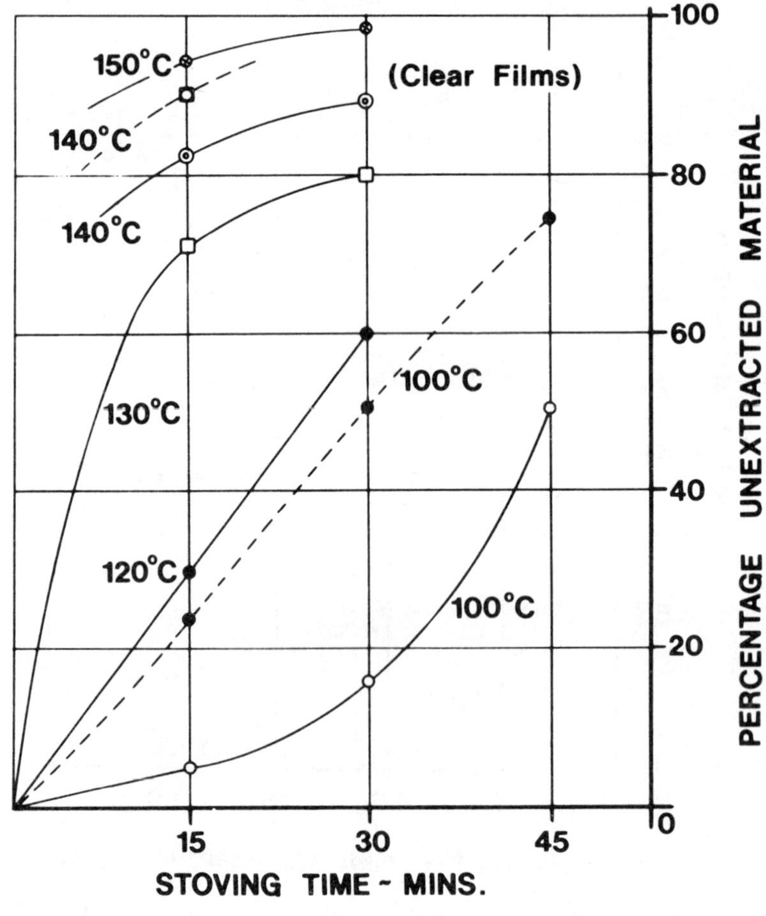

Fig. 11 Curing rates of acrylic – low and high temperature bakes

B represents a reflow thermosetting acrylic resin which was designed to pass the Ford reflow specification ESB.M32J.102A. Both resins have fairly high solids contents of about 60% and viscosities of 15 to 30 poise at 25°C.

A white enamel base on resin A reflowed satisfactorily after an initial low temperature bake of 15 to 30 minutes at 80°C or 5 to 10 minutes at 90°C (oven temperature) but was too highly crosslinked to reflow after an initial bake of 15 minutes at 100°C (panel temperature).

Resin B reflowed satisfactorily after stoving for 15 minutes at 100°C (panel temperature) as a clear film and when pigmented, although it was found that both acrylic melamine ratio and the pigmentation had a profound effect on its reflow properties (Fig.11).

Sandability

Assessment of dry sandability of a film after the initial low bake presented some difficulties, especially in a series of enamels, since the results could not be duplicated with the degree of accuracy required, especially when comparing results between two operators. An attempt was therefore made to correlate sandability with film hardness as measured by (a) the pencil method, and (b) the Vickers Diamond Indentation apparatus.

Metallic finishes

The increasing use of metallic finishes in automotive enamels is well known and is possibly the reason for the widespread acceptance of hydroxylated acrylic finishes which technically provide an ideal carrier for non-leafing grades of aluminium pigment. Metallic enamels based on acrylic resins possess excellent gloss, colour retention, and resistance to acid spotting, and good durability. In these respects acrylic melamine systems are usually more acceptable than alkyd melamine in metallic finishes.

Reflow metallic enamels

The effect of addition of aluminium pigments to a reflow acrylic resin system was examined. Particular note was made of the position of the aluminium flakes in the film before and after reflow. It was quite evident that if the pigment particles were not uniformly dispersed throughout the thickness of the film, colour changes would take place during reflow.

Conclusions

The performance of hydroxylated acrylic melamine resins in reflow enamels is dependent upon several interrelated factors, including the chemical constitution of the resins, type and pigmentation, and the ratio of acrylic melamine resins in the paint. It has been indicated that it is possible to formulate a resin having the required speed of cure to give satisfactory sanding, reflow and overspray acceptance properties for repair in process operations, and at the same time to be resistant to cold checking under severe conditions in thick films. In view of the possible legislation regarding air pollution, modifications of existing acrylic polymers and continued resin development appear to be necessary. The weathering properties of thermosetting acrylic resins under severe conditions, especially when used in metallic enamels, have been shown to be extremely satisfactory.

The research and development that has taken place over the last twenty years is quite remarkable. All the major paint and chemical companies have expended millions of pounds to develop the technology behind every tin of automotive material. There is no sign that this great advance will slow and the motor manufacturer and the refinisher can look forward to new and exciting developments.

Manufacturers' instructions

To gain the maximum performance from all the various paint types it is necessary always to follow the paint manufacturer's instructions. So often, misguided painters believe that they can improve on the procedures set out by the manufacturers, but this is not so. The use of non-standard or 'cheap' thinners is a case in point. Set out below is the formulation for a cellulose material showing the thinner balance. Unless a substitute thinner matches the formulation of the thinner designed for the colour then a sub-standard finish must result. It is a very shortsighted policy indeed to break away from the paint manufacturer's recommendations.

Cellulose formula

Nitro-synthetic formulaion

½ sec nitrocellulose (spirit damped)	9.7%
15–20 sec nitrocel (spirit damped)	0.5%
Medium oil length non-drying alkyd at 50% solids	12.6%
Rutile titanium oxide	4.8%
Solvent	72.4%
	100.0%

Solvent blend

MIBK (methyl isobutyl ketone)	10%
Isopropanol	15%
Cellosolve	15%
MEK (Methyl Ethyl Ketone)	30%
Xylene	20%
Toluene	10%
	100%

It is important to stress that all recommendations of flash off times, drying times, number of applied coats, etc., have been carefully checked again and again by the manufacturers so that they are sure that their product gives the maximum performance to the operator.

All the major paint producers' materials are of the highest standard and, used correctly, will give excellent results time and time again. It is extremely rare that a fault experienced after the application of materials is due to the manufacture of the paint.

Powder coatings

In recent years great strides have been made in powder coating technology and many motor manufacturers are now using these materials.

The history of powder development and Arthur Holden & Sons Ltd's involvement go hand in hand. The company commenced investigation into the processes of powder coating in 1963 and its commitment has continued ever since with considerable investment in both development and manufacture.

The first powders, which were based on epoxy resin, were applied at high air pressures to ensure a full cloud; film thicknesses were around 100–125 microns and stoving was based on 200°C for 30 minutes. Considerable progress has been made since then with both powders and equipment. Standard materials can now be applied with film thicknesses as low as 35 microns, cured at 160°C article temperature for 10 minutes. Powder emission rates are much lower and transfer efficiency greater, so that far less overspray has to be collected and recycled.

Resin manufacturers have played their part in the development of powder coating. Resins and curative systems have become more sophisticated and new resin types have become available. To a considerable extent the mixed polymer system epoxy polyester has replaced epoxy as the general purpose powder.

Neither epoxy polyester nor epoxy can be recommended for outdoor applications where chalking cannot be tolerated, although the protection afforded by the film will be unimpaired. For outdoor durability polyester is normally recommended.

Certainly suppliers of small parts to the motor industry are coating up with powder because of the advantages, and it is worth noting them:

- Powder usage normally implies a one-coat system. For specialised purposes, occasionally it may form part or the whole of a two-coat system.
- Overspray may be collected, sieved and recycled for further use.
- Powder is used straight from the container in which it is supplied – no viscosity checks, thinning, meter readings, as are necessary with wet paint.
- Mechanical properties are excellent, resulting in far less damage when piece parts are handled and assembled.
- Powder coatings may be machined.
- Powder is easy to apply by hand, an unskilled operator soon establishing the correct technique.
- Powders lend themselves to use through automatic spray equipment with ease of control of film build.
- Fire and explosion risks are reduced as compared with solvent-borne paints.
- Virtually no volatiles are involved, and therefore less ventilation is required.

- No flash off zone is required between spray section and stoving plant – hence saving in space.
- Powder plants, particularly automatic plants, occupy less space than wet paint systems.
- Properly designed powder application systems offer considerable energy saving compared with most wet paint systems.
- Correctly formulated powders have good edge coverage and do not 'run'.
- Plant maintenance is easy. The basic requirement for cleaning the recovery booth is a suitable vacuum cleaner.

The actual manufacture of powder coating is carefully controlled and resin, pigments, hardeners and additives such as flow agent are loaded into a blending machine which mixes the raw materials for a determined time to obtain a homogeneous mix. Samples are taken from the dry blend mix, extruded on small-scale equipment and checked for colour (using a colour computer), gloss and appearance against the master pattern. If incorrect, the necessary additions are made, the blend is remixed and further checks made. It is imperative that the mix be correct at this stage as, once extruded, no further adjustment can be made. When correct, the dry blend mix passes on to the next stage – extrusion.

The dry blend mix is fed to the extruder, the barrel of which is heated to a temperature allowing the mix to become molten. The screws of the extruder, when revolving, cause 'shear' forces to be exerted on the premix which disperse the pigments and additives into the molten resin system. The hot mixture emerges from the die head on the extruder to be passed through a cooling system, usually water-cooled rollers, followed by further cooling on a conveyor.

Extrusion temperature varies from 80°C to 130°C depending on the type of system being manufactured and extrusion conditions are carefully controlled to ensure that no premature reaction takes place. Cooled extrudate, which is brittle at ambient temperature, passes through a kibbler (or crushing machine) to produce pieces approximately ¼ inch square.

Kibbled extrudate is passed through a grinding machine to produce a finely ground powder of a predetermined particle size distribution. Granulometry can be adjusted to suit the requirements of a particular application equipment. The final operation of sieving removes any oversize particles.

At all stages of the manufacturing process cleanliness is important to avoid contamination, whether of one colour with another or between non-compatible resin systems. Cleaning is a time-consuming operation which can be minimised by careful production planning, but not eliminated, and downtime remains a major item of production expense. Since downtime is the same for small and large batches, a minimum batch size has to be stipulated.

Preparing the parts for coating is as important as the application. The requirement for successful powder coating is that powder be applied to clean, dry, grease-free metal. A coating which appears well-adhered initially may break down after a very short time if the substrate pretreatment was inadequate.

The presence of oils and greases, whether deliberately applied as a temporary protection or carried over from a previous engineering operation, can be readily detected; so too can mill scale and rust. More troublesome because they are both less apparent and more difficult to remove are mould lubricants on castings and die lubricants on extruded sections. The surface of extruded aluminium sections has, in addition, a film containing magnesium oxide which must be removed.

In addition to cleaning, positive pretreatment may be required to enhance adhesion and increase long-term corrosion protection. Chemical pretreatment of ferrous, and chromate pretreatment of aluminium substrates inhibits underfilm corrosion creep and a well applied pretreatment/powder coating system will give satisfactory performance for many years. Extreme examples are bus chassis, for which epoxy powder on zinc phosphated hot-dipped galvanised steel is specified, and aluminium window extrusions which use a system of polyester powder over chromate pretreatment.

This is a simple method of cleaning, but prone to misuse. Dirty rags and dirty solvent can put back almost as much contaminant as they take off. Clean rags, plus solvent used for a limited period then discarded, must be the order of the day.

Vapour degreasing using trichloroethane (which is replacing trichlorethulene on health and safety grounds) is a much favoured system for removing oils and greases, but it will not remove soils adequately and a further wipe over with a clean cloth may be necessary. This disadvantage may be overcome by utilising the materials in a pressure jet system.

A wide range of alkali cleaners are available to remove most oils, greases and soils even up to heavy concentrations. Special grain refining agents can be included to promote specific phosphate coating properties. The workpiece may be either dipped or sprayed with the hot aqueous cleaner but must be thoroughly rinsed, preferably twice, before passing on to the next stage of the pretreatment.

A system utilising sulphuric and/or hydrochloric acid will remove rust and light scale. The workpiece must be rinsed thoroughly before the next stage of the pretreatment.

Ultrasonic agitation can be allied with various cleaning methods to assist in the removal of soils. This system will remove rust and mill scale but incorrect choice of shot can result in a surface profile on the substrate which is unsuitable for powder coating. If powder is to be applied at, say, 50 microns, profiles should not be above 25–30 microns. GO7 and G12 grits have been found to be suitable.

After shotblasting, surface dust must be removed prior to the next stage which must take place without delay as freshly shotblasted surfaces rapidly oxidise.

After thorough cleaning, the workpiece should pass immediately to the phosphating stage and the norm is an orderly progression through a series of tanks or spray stations. Different phosphate systems are available for ferrous, zinc coated, mixed ferrous/aluminium systems and the advice of the pretreatment supplier should be taken on the one most suitable for the application.

In recent years low temperature phosphating systems, as outlined above, have been introduced, with process temperature reduced from 80–90°C to

35–40°C as an energy saving measure.

Duraplast powder coatings are formulated for application by electrostatic spray. Application may be by a single hand gun or one or more guns forming an automatic set-up. Powder from the feed hopper is conveyed on an airstream to the gun where it is given an electrostatic charge. The charged particles are attracted to the earthed workpiece. Powder adheres through electrostatic attraction and is difficult to remove by normal air movement or jolting. Figure 12 illustrates the layout of a powder application system.

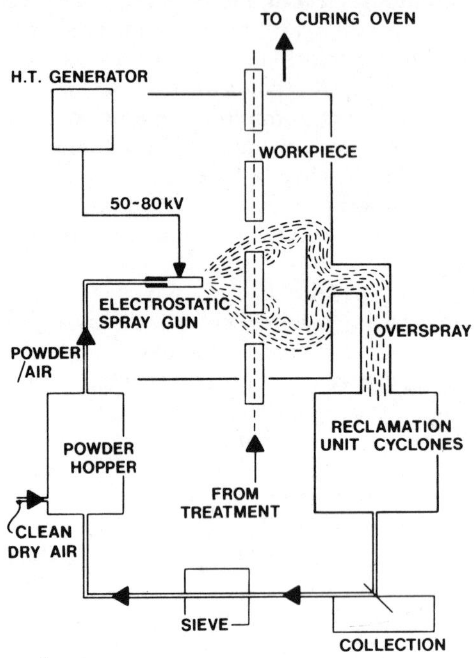

Fig. 12 Powder application system

Powder emission rates, air pressures and voltage settings are normally established at the plant commissioning stage in close collaboration with equipment and powder suppliers. Operating experience and changing work loads may necessitate adjustments, but constant 'fiddling' with the controls should not normally be involved.

Workpieces can be inspected prior to stoving and any bare or damaged areas touched in with powder. In the last resort, the unstoved powder can be blown off with compressed air and the article recoated. The ease and convenience with which faults can be rectified at an early stage is a factor which contributes to the very low reject rate of powder coating finishes compared with wet paint.

After coating, the article moves directly to an oven (no 'flash off' required for powder) where the powder melts, flows and cures. Recommended cure schedules are normally quoted as, for example, 160°C article temperature for 10 min. (Fig.13). (Allowance must be made for the time necessary for the article to reach the specified temperature.)

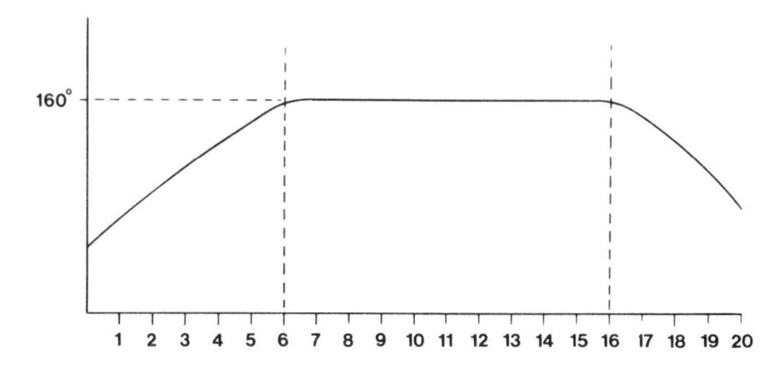

Fig. 13 Temperature profile of powder coating

6

Anti-rust

A great deal has been written and discussed with regard to rust-proofing and protection. It is certain that whatever compound wax injection, lanolin treatment or plain underseal the manufacturer uses, it really only holds back for a longer period of time the inevitable rusting process. Due to the constant damp in Northern Europe and the heavy use of salt on the roads, the onset of rusting and corrosion is only a matter of time.

Although many motor manufacturers are offering long and extended warranties on paint and body, it has yet to be proved how these claims of rust-proofing will stand the test of time and although technology in these fields has improved, mass-produced cars are being made of lighter gauge steel which will rust through that much more quickly.

Rusting in cosmetic panels is caused by breakdown in the paint film, i.e. stone chips or a scratch, or by osmosis (see later in this chapter).

It appears that once rust has attacked a panel and established even a small foothold, then total rust out will occur.

Of the many companies active in rust prevention, one is certainly worthy of consideration. Mebon Paints Ltd make and market a material known as Coachguard. I have examined panels which have been coated in this material and then exposed to prohesion tests (tabled below) that give convincing evidence of rust protection. This company is enjoying well-deserved success in the coach building industry. With products like these, which are easy to use and relatively inexpensive, progress is being made to hold back the inevitable rusting process.

The fact is that every car, from a Rolls-Royce to a Mini, has a life and eventually its life in that form must end. Only persistent rebuilding and panel replacement will ensure the vehicle can continue in its prime function, that is, to carry passengers comfortably and in safety.

Coachguard protective system – Mebon test report

Scope

In order to establish a method of laboratory testing which would effectively reproduce vehicle operating conditions, Mebon decided to subject the Coachguard system to a series of British Standard tests renowned for their speedy destruction of lesser protective systems. The test was carried out in three stages:

A. Corrosion resistance.
B. Resistance to mechanical damage.
C. Resistance to attack from chemicals normally associated with road vehicles.

The findings listed below were derived not only from Mebon's own laboratories but from tests carried out using various independent motor industry research facilities.

Findings

A. *Corrosion resistance*

Test	*Duration*	*Result*
(i) Mebon prohesion test	500 cycles/1,500 hours	Pass
(ii) Salt spray	ASTM B117 1,000 hours	Pass
(iii) Humidity BS 39000/F2	1,500 hours	Pass
(iv) Artificial weathering BS 3900/F3	2,000 hours	Pass

The above tests simulate the effects of road spray and road salt corrosion. A pass is determined as no visible signs of breakdown.

B. *Resistance to mechanical damage*

Test	*Result*
(i) Impact test BS 3900/E3 – no detachment reverse and direct impact	Pass
(ii) Chip resistance B5 AU148 – no exposure of substrate	Pass
(iii) Chip resistance + salt spray – B(ii) + A(ii) (250 hours)	Pass
(iv) Bend test BS3900/El – $\frac{1}{4}$ inch mandrel	Pass
(v) Adhesion BS390/E6 – no detachment (Class 1)	Pass
(vi) Scratch Test BS3900/E2 – 2,000 grams maximum	Pass

The above tests simulate the combined effects of mechanical damage, i.e. stone chipping, abrasion, etc.

Prohesion testing

Panels	Box sections mild steel
Panel source	Hestair Dennis via S. Heidukewitsch
Panel reference	J12A J12B
Exposure date	30.1.1984
Removal date	12.3.84
Examination date	12.3.84
Exposure time	42 days = 1,000 hours
Temperature during spraying	Approximately 22°C
Temperature during drying	35°C
Spray solution composition	0.5 gm/litre sodium chloride, 3.5 gm/litre ammonium sulphate
Cycling frequency	1 hour spraying, 2 hours drying
Approximate outdoor time equivalent	Approximately 5 years

Film thickness
The 'A' panel was coated with Coachprime-Coachguard to a total average DFT of 120 microns. The 'B' panel was Coachguard only to 50 microns average dry film thickness.

Performance
The 'A' panel generally performed very well. There was up to 5 mm undercut corrosion in limited areas on a cross that had been scribed on the panel. There were no other signs of corrosion or breakdown anywhere on the panel.

The 'B' panel performed far below the standards of 'A' panel. There was minimal undercut corrosion on the scribed cross, but corrosion was evident, and film breakdown also evident, throughout large areas of the panel.

Conclusion
The system on panel 'A' (Coachguard-Coachprime) was applied at approximately the correct film thicknesses and gave the expected good performance.

The system on 'B' (Coachguard) was under the specified film thickness by approximately 50%. This resulted in a performance which was extremely poor.

C. *Resistance to attack from chemicals normally associated with road vehicles*

Test		*Requirement*	*Result*
(i)	Engine oil resistance	2 hours immersion @ 80°C + 1 hour recovery – no deterioration	Pass
(ii)	Gear oil resistance	As C(i)	Pass
(iii)	Antifreeze resistance	BS3900/G3 33% antifreeze @ 50°C	Pass
(iv)	Battery electrolyte	BS3900/G3 sulphuric acid – no spotting after 24 hours	Pass
(v)	Diesel fuel resistance	BS3900/G1 24 hours at ambient – as C(i)	Pass
(vi)	Burn rate	2 minute maximum	Pass
(vii)	Steam cleaning	10 washes – no deterioration	Pass

The above tests simulate the various chemical attacks associated with road vehicles. All tests were conducted on panels prepared in accordance with BS3900 and coated with 50 microns dry film thickness Coachprime 4-1-28 and 100 microns Coachguard 4-2-28.

Wax injection and the injection of lanolin-based materials into box sections are now part of the rust-proofing exercise carried out by all the manufacturers. This obviously helps to delay the rusting out and it is important to re-proof vehicles after accident repair. Often after repair, where nitrocellulose-based materials are used they appear to excite the original finish and rusting in the cosmetic panels occurs very quickly. This does not appear to be so apparent when two-pack isocyanate products are used. Any high build material with polyurethane bases seems to be more dominant in this area.

Osmosis

The reason why rust begins in the centre panels of a motor vehicle is due to osmosis; moisture comes from the outside of the vehicle and penetrates the paint film by osmotic pressure. Phosphate mineral salts are deposited in the grain boundary of the metal during the manufacturing paint process. The vehicle is painted in its full primer and colour coats, sealing in the phosphate deposit. On exposure to the atmosphere there is a migration of moisture that strikes at these pockets of phosphate. This occurs due to the fact that all paint is a semi-permeable membrane. The moisture becomes a salt solution and then by osmotic pressure more moisture is drawn into the film. Through the film the weaker moisture flows to the stronger salt solution until at last the rust blemish causes an upheaval that breaks the surface. When this spot of rust does break the film surface, it will be found that underneath a considerable area of rust is present.

When this occurs throughout a panel it must be stripped, and usually these spots of rust have joined up under the film by capillary action. The remedy after stripping the panel to bare metal is to abrade the whole panel with a DA sander using an 80 grit disc. The rust pockets must be entirely removed if it is not to occur again. Metal treatment with phosphoric acid is necessary to clean the panel and prepare it for self-etch primer. The panel can then be finished in the normal way.

7

Basic Preparation

The key to total success in fine finishing lies in the basic preparation of the vehicle.

As I am only concerned in this book with fine finishing, stripping to bare metal is essential within the process (Fig.14). An old system, being subject to amino acids attack, saline solution and the general road film grime, will have broken down, and it is not reasonable for that to be the sound substrate on which to build a fine finish that will last.

Fig. 14 A rare Aston Martin DB6 estate car stripped to bare aluminium awaiting cleaning and priming in self-etch primer. Note the lights and brightwork refitted for check purposes prior to removal before repainting

The following process will give a finish that will last and enhance the vehicle for years to come.

1. Strip all bright parts from the vehicle, bumpers, lights and trims of every kind. Mask up all glass.
2. Using Meltic or Nitromors paint stripper, remove the total paint film to bare metal.
3. Wash vehicle in hot water and then flat with 180 wet and dry paper.
4. Dry vehicle and examine. Any dents should be panel beaten and filed. If this is not possible then bodyfiller or lead should be used.
5. Vehicle should be treated with ICI Deoxidene. These products are

phosphoric acids in solution and will clean out the grain boundary of the metal (steel or aluminium). They will leave the metal open for the following process. Ensure that the material is washed off with hot water and is never allowed to dry on, as phosphate salts can occur. When the vehicle is completely dry, examine it carefully but do not touch. It is important that any polyester body filler has not been contaminated with the phosphoric acid and normally any area where this is present is masked out.

6. Replace masking on vehicle and air dust down. Apply by spray a single coat of self-etch primer. This material is a zinc chromate base mixed with orthophosphoric acid. The acid eats into the metal and carries with it the zinc chromate. It is essential that it is used on GRP or aluminium vehicles. Without it adhesion will certainly break down. As soon as this has dried it must be coated up.
7. Either two-pack filler or nitrocellulose-based primer filler can be applied. Follow the manufacturer's specification with special care when thinning to the correct viscosity for spray application. Normally two or three coats are applied. This material can be low baked at 80°C panel temperature for 15 minutes or longer depending on type of material. It can be left to dry overnight. Unless conditions are controlled an uptake of moisture can occur causing micro blistering in the colour film later.
8. Black guide coat must then be sprayed all over the vehicle. This is wet flatted using 600 wet/dry paper. It is preferable to use a rubber or wood rubbing block to ensure that a perfect surface finish is obtained. When this operation is complete then a rinse down and careful examination of the film is necessary.
9. At this point either a sealer coat or a further coat of primer filler can be applied.
10. After thorough drying, colour coats can be applied. To give sophistication to the paint finish, either a full nitrocellulose finish or a thermoplastic acrylic must be used. Two-pack isocyanate materials tend to be full or 'fatty' in appearance and at present do not give the clarity of film appearance.

It is necessary to obtain full cure out before flatting and polishing.

If the material is low bake then after the full cycle the vehicle cools to ambient air temperature. In air drying materials a time span of at least 48 hours should elapse before the flatting procedure.

Flatting should be carried out with 1200 wet and dry paper with the aid of soap. This acts as a lubricant and helps to flat the film back without deep scratching or clogging. Use a block at the flat of the hand and work until the paint film is totally matt. Great care should be exercised in this operation as it is very easy to both 'furrow' the paint film and leave flatting marks, and break through on edges. This obviously means repainting the area. With straight colours this is relatively simple but with metallic and basecoat and clear finish systems the problems of making an 'invisible repair' are much more difficult. The art of 'fading in' areas is dealt with separately in Chapter 12, on Rectifications.

Having flatted the vehicle carefully from end to end, the cutting back of the film then begins. A fine cutting compound is used and applied with soft mutton cloth. This is hard work and the operation will take hours of hard rubbing to cut the surface back to a finer finish. When this has been achieved the next stage is to cut the surface again with a fine liquid solution such as 'T' – cut' made by Tetrosyl. This will lift the finish to a further high gloss. It may be necessary to cut the vehicle more than once with this material.

Finally, to take out the fine scratch marks left by the mutton cloth, the vehicle should be wax polished. Certain polishes have a heavy silicone content as a slip additive, but the presence of this material causes a contamination that is difficult to remove when refinishing the vehicle again.

The vehicle should be beginning to look very presentable and the fitting up procedure can now begin.

8

Application

The art of a fine finish rests not only on the hours of preparataion that have been completed but also on the actual application of primers, fillers, sealers and colour coats.

The whole idea of spray application is to convey the paint to the panel and give a coating of even depth throughout and totally encapsulate the panel in a paint system.

The gun pass speed, and the distance from the panel, are important aspects, along with all the other factors such as spray booth temperature, air movement, size of the vehicle and physical build of the sprayer, which is a point often overlooked. It is helpful to be tall and have a good natural rhythm to body movement.

A good sprayer should be on his toes like a boxer or dancer and move gracefully and under perfect control. Every movement carried out will be reflected in the film build of the material and if this is inconsistent then eventually paint breakdown will occur in patches where least paint has been applied. Smoothness and rhythm are the key to fine finishing.

Basics in application

- The gun should be held at right angles to the panel.
- The gun should be 6 to 8 inches from the panel.
- There should be a 50% overlap for primers, fillers and straight colours and up to 80% for metallics that tend to 'band' or 'float'.
- At the end of every stroke there should be a trigger off and restart.
- Do not arc the gun as this will cause dry spray and a change of pattern in metallic.
- Plan the job and on a full respray start with the roof and work down and round the car.
- It is important to remove both the bonnet and boot of the vehicle to ensure an accurate job, and to prevent dry overspray contaminating these panels when finishing the roof.
- Keep paint application on the light side so as to avoid solvent trapping and excessive film builds. These will only shrink back in time, giving sinkage areas over repairs and spoiling the job.
- It is better to coat up a vehicle and allow it to through dry, then flat back and recoat, than to keep applying paint and hope to flat and polish later. The film remains soft and solvent trapping is almost certain to occur with excessive build.
- Keep the gun clean, and the sprayer must wear overalls, mask and cap to ensure safety and cleanliness of the job.

Start with the roof of the vehicle and, working from the centre, spray back to the gutter. Pick up the wet edge on the roof from the other side and overlap to keep the join wet and proceed to the other gutter and then come down the windscreen pillars, the B post and the rear panel to the boot level. Work round the vehicle from the rear panel to the front and spray down each side, keeping wet edges all the time.

If you finish back at the rear panel (roof to boot) the overspray will be contained in one small area for flatting and polishing.

Colour coats

Often a single wet coat, to set up the paint film and allow solvent penetration of the filler coat, can be followed by a wet on wet. This is when colour is applied from top to bottom of the panel and immediately recoated in the same manner. This gives high build and superb 'flow out' and gloss. However, there is a danger of over-application that will result in runs and sags that have to be rectified later.

Let us assume that an average-size air compressing plant of about 12 c.f.m. free air is being used and that a type JGA spray gun fitted with a size 30 air cap and a size EX fluid tip and fluid needle is being operated. To commence operations the spray painter will connect the air hose leading from the air outlet on the transformer to the air inlet on the spray gun.

Next, the paint will be reduced to the proper consistency as recommended by the paint manufacturer, thoroughly mixed and strained through gauze or mutton cloth into the suction feed cup or the remote pressure feed cup, which is then attached to the fluid inlet of the spraygun. The air inlet valve on the transformer is then opened and the air adjusted by means of an air adjustment valve on the transformer to the required pressure, in this case 55 lb per square inch. The gun is then adjusted for spraying in the following way: the spreader adjustment valve is opened as far as it will go, and the fluid adjusting screw is turned in an anti-clockwise direction until the first thread of the screw becomes visible.

Spray pattern

A test pattern is then sprayed onto waste material to check for uniform paint distribution. This is done by holding the gun steady and momentarily triggering off and on. A normal pattern for this type of gun should be approximately 10 inches wide when the gun is held 6 to 8 inches from the surface. Then a small area is sprayed to check the speed of operation. If the spray pattern appears starved of material the fluid adjusting screw is opened wider to allow more paint through. If too much paint is applied the paint flow is reduced by screwing in the fluid adjusting screw or by reducing the air pressure at the transformer. If the remote cup is used the fluid pressure is reduced by means of an adjusting valve in the cup lid.

If the atomisation is too fine (this is recognised by a speckled effect or dimple finish lacking flow out of paint), the atomising air pressure is increased or the material flow cut down.

If, when the spreader valve and fluid screw are wide open, the spray pattern is too narrow, the fluid is increased by raising atomising pressure or by thinning the material.

A wetter coat may be obtained by turning in the spreader valve to narrow the spray pattern and then slightly increasing the atomising pressure.

With a gravity feed spraygun the action taken is exactly as above and the size of set-up is the same also.

Technique

Spraygun technique is probably the most important single item contributing to a successful paint job – a tool is only as good as the technique or 'know-how' of the operator who handles it. This is as important and true of a spraygun as of other tools. Good spraygun technique involves four principal rules:

- Gun distance – 6 to 8 inches from surface.
- Keeping gun at 90° (perpendicular) to surface.
- Overlap each spray stroke 50 per cent.
- 'Trigger' at the end of each stroke.

Let us take a closer look at each of these points.

Gun distance

In order to obtain the best results from the spray process we must take full advantage of the manner in which the gun was designed to operate. An air cap, which is the forwardmost part of the gun, is designed to give its best performance within a range of 6 to 8 inches from the surface being sprayed. At

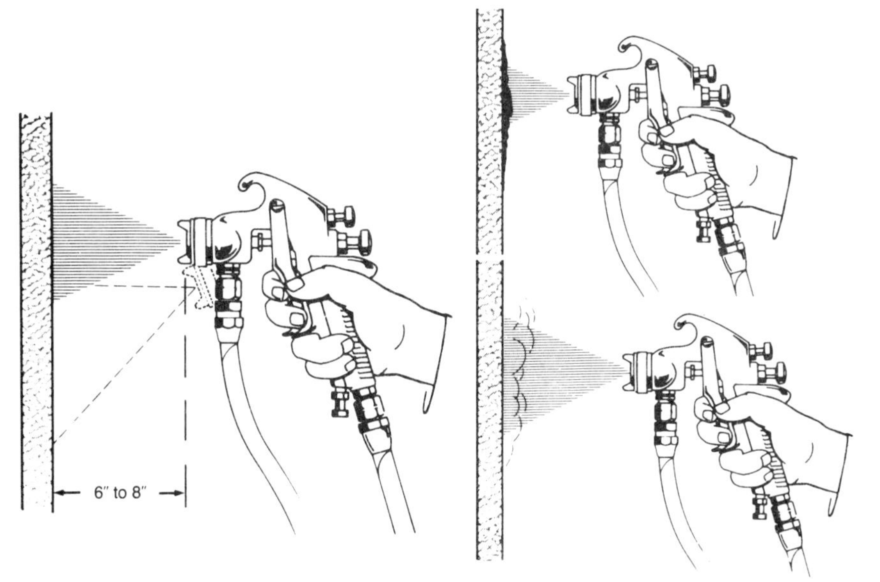

Fig. 15 Correct spraygun distance **Fig. 16** Incorrect spraygun distance

this distance it will create its best atomisation and distribute the fluid coming out of the gun to the full width of the pattern size (Fig.15).

If the gun is operated any closer to the surface, it will not allow the air cap to achieve good atomisation and distribution. The coating will be in a very wet, non-uniform sagging condition (Fig.16). If the gun is held at a greater distance from the work than recommended the result will be a dry, sandy, uneven coat of paint. At excessive gun distances there will be an increase in overspray and when this occurs a greater amount of paint will be blown away and lost (see Fig.16).

At all times when the gun is moved back and forth or up and down across the surface this 6 to 8 inch distance must be maintained.

Gun perpendicular to surface

To ensure a uniform deposit of paint on the surface it is imperative that the gun be stroked across the surface at a right angle to it. Pointing the gun to either side (arcing) or up and down (angling) will result in a non-uniform deposit of paint giving an uneven spray pattern. The wrist should be kept flexible during the spraying process (Fig.17).

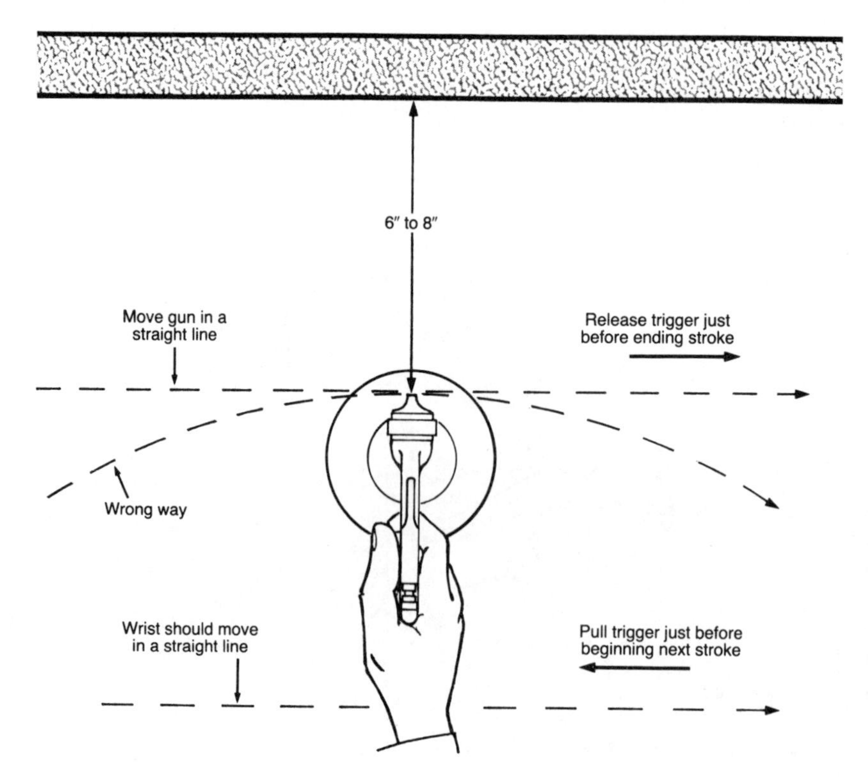

Fig. 17 Arcing of the spraygun

In certain types of work it is often necessary to tilt the gun but this should not be done on surfaces suited to the correct gun position. Normally, the stroke is made in a horizontal direction to the object being sprayed. If a vertical stroke is preferred or is necessary the operator need only loosen the air cap retaining ring and rotate the air cap 90°. This gives a horizontal fan pattern and will facilitate vertical painting.

Overlap

As the gun is moved back and forth across the surface it is important that the spray be directed so as to overlap the previous strokes by a full 50 per cent. This eliminates the need for a double or cross coat and will assure the deposit of a wet coat without the risk of streaks or patches. Long work is sprayed in sections of convenient lengths, each section overlapping the previous section by 4 inches (Fig.18).

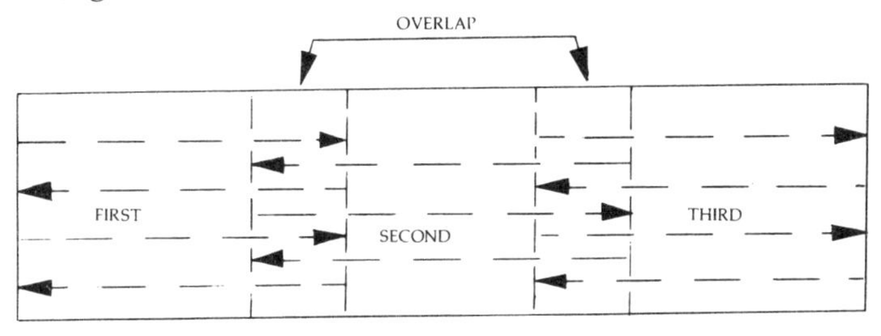

Fig. 18 Spraygun overlap

Triggering

At the end of each stroke the gun should be 'triggered' before starting the next stroke. This is done by releasing the trigger so that the flow of paint will be momentarily stopped while moving the gun to position for the return stroke. It is only necessary that the flow of paint be stopped. Upon developing a feel or touch for the trigger it is accepted practice to permit the flow of air through the gun at all times, stopping only the paint. This also allows a slight increase in speed. Failure to 'trigger' the gun will result in an increase of overspray and may cause sags, runs, and streaks.

Positioning

When spray painting it is recommended that the operator position himself to the 'left of centre' of the item being painted. This enables him to reach and bend freely without violating the foregoing rules of spraying technique.

Cleaning and lubricating the spraygun

A *suction feed* spraygun and cup should be cleaned as follows:

- Loosen cup from the gun and while the fluid tube is still in the cup, unscrew

the air cap about 2–3 turns, hold a cloth over the air cap and pull the trigger. Air is thereby diverted into the fluid passages forcing the material back into the cup.
- Empty cup of material and replace with a small quantity of solvent. Spray the solvent through the gun to flush out the fluid passages.
- Remove the air cap, immerse it in clean solvent and dry it by blowing off with compressed air. If the small holes become clogged soak the cap in solvent. If reaming should be necessary, use a matchstick, broom bristle or orange stick. Do not dig out the holes with wire or nails as this may permanently damage the cap and affect the spray pattern.
- After the cleaning of the air cap replace it on the spraygun, wipe the gun with a solvent-soaked rag and if necessary brush the air cap and gun with a fibre brush using cleaning fluid or thinner.

A *gravity feed* gun is cleaned as follows:

- Remove the cup lid, empty out any paint remaining in the cup and quarter fill it with solvent or cleaning fluid. Do not replace the cup lid: unscrew the air cap ring two or three turns, hold a cloth over the air cap and, with the cup opening turned away from the face, pulle the trigger causing the air to bubble back into the cup.
- Remove and replace the cloth on the cap several times keeping the trigger pulled and finally spray the remaining fluid through the gun.
- Clean air cap and fluid tip and wipe the gun, cup and lid with solvent-soaked rag.

A *pressure feed* system is cleaned as follows:

- Shut off the compressed air from the tank by means of the pressure regulating valve on the lid.
- Release the pressure in the tank by opening the relief valve and loosen the cover.
- Unscrew the spraygun cap ring three turns, hold a cloth over the air cap and pull the trigger, forcing material in the gun and hose back into the tank.
- When the hose is empty detach it from the tank and place the free end in a small container filled with suitable cleaning fluid. Operate the gun in the usual manner holding it as near to floor level as possible. The fluid is thus siphoned through the hose and gun, cleaning them thoroughly.
- Dry out hose with compressed air.
- Clean air cap and fluid tip as before.
- Clean out tank and re-assemble for future use.

Note: It is a common malpractice to clean sprayguns by placing the entire gun in solvent. This can cause the air passages of the gun to be fouled with sludge which may subsequently be blown onto a surface being sprayed. Solvents also remove lubricants and dry up packings.

Do not use caustic alkaline solutions to clean sprayguns because they affect the aluminium alloys used in gun bodies and parts.

Lubrication

After each cleaning of the spraygun, place a drop of oil on the fluid needle packing, the air valve packing and the trigger bearing screw. The fluid needle packing should be removed occasionally and softened with oil and the fluid needle spring should be coated with Vaseline.

The parts of the gun which require lubrication are the fluid needle packing, air valve packing, the trigger bearing screw, and the fluid needle spring.

Spraygun troubles and remedies

Leakage from the fluid needle packing nut

This is caused by a loose packing nut or worn or dry fluid needle packing. Lubricate the packing with a few drops of light oil. Tighten packing nut to prevent leakage but not so tight as to grip the fluid needle. Replace packing if unduly worn.

Air leakage from the front of the gun

This can be caused by:

- Foreign matter on valve or seat.
- Worn or damaged valve or seat.
- Broken air valve spring.
- Sticking valve stem due to lack of lubrication.
- Bent valve stem.
- Packing nut too tight.
- Gasket damaged or omitted.

The remedy for all these causes is obvious.

Fluid leakage from the front of the gun

This is caused by the fluid needle not seating properly due to:

- Worn or damaged fluid tip or needle.
- Lumps of paint or dirt lodged in fluid tip.
- Packing nut too tight.
- Broken fluid needle spring.
- Wrong size needle.

Again the remedies are obvious.

Jerky or fluttering spray

This is caused by air leakage into the fluid line and is due to:

- Lack of sufficient paint in container.
- Tipping container at excessive angle.
- Obstructed fluid passageway.
- Loose or cracked fluid tube in cup or tank.

- Loose fluid tip or damaged tip seat.
- Too heavy a paint for suction feed.
- Clogged air vent in cup lid.
- Loose, dirty or damaged coupling nut or cup lid.
- Dry packing or loose fluid needle packing nut.
- Fluid tube resting on bottom of cup.

To remedy the last trouble, carefully bend the fluid tube slightly upwards to clear the bottom of the cup.

Defective spray pattern

If dirt or dry paint become plugged into the horn holes of the air cap, on the top of the fluid tip or on the air cap or fluid tip seat, a defective spray pattern will be obtained and this should be rectified by careful cleaning. A defective spray pattern will show when a test trigger pull is made and the shape of the pattern indicates where to look for the obstruction. The pattern can be top heavy, bottom heavy or side heavy and the position of the obstruction can be determined by rotating the cap one half-turn and spraying another pattern. If the defect is thus inverted the obstruction is on the air cap. If not inverted it is on the fluid tip. Whilst the air cap is easily cleaned as previously instructed the defective pattern may be caused by a fine burr on the edge of the fluid tip or by dried paint just inside the opening. In the former case it can be removed with 600 wet or dry sand paper, and in the latter by cleaning.

Sometimes a heavy centre pattern is obtained and this is caused by:

- Too low a setting of the spreader adjustment valve.
- Too low an atomising pressure or the paint being too thick.
- Too large a nozzle for the paint used.
- Too small a nozzle.

The remedy is to readjust the atomising air pressure, fluid pressure and spray width control until the correct spray pattern is obtained.

Sometimes when pressure feed is used a spray pattern is split due to the air and fluid pressures not being properly adjusted. Reduce the width of the spray pattern by means of the spreader adjustment valve or increase fluid pressure, adjusting atomisation pressure as necessary. It should be remembered that this latter adjustment increases speed and the gun must be handled much faster.

Orange peel

Orange peel is a finishing defect common to cellulose and synthetic materials, so called because it resembles the texture and the appearance of orange peel. General causes are:

- Wrong thinner.
- Atomisation pressure either too high or too low.
- Gun held either too far away from or too close to the work.
- Material not thoroughly mixed or agitated.
- Draught in the finishing room.

- Improperly prepared surface.
- When spraying synthetics, too low humidity.

Streaks on finished surface

These are caused by:

- Tilting the gun. One side of the pattern hits the surface from a shorter distance, causing more material to be applied at this point.
- Air cap or fluid tip may have dirt or burrs on it, causing heavy top or bottom patterns.
- Split spray causing more material to be applied at the top and bottom of the pattern.
- Spray patterns not properly overlapped.

Runs and sags in finishes

These are the result of too much material being applied to the surface, and also the material being too thin. If the gun is tilted at an angle, excessive material is applied where the pattern is closest to the surface causing it to pile up and sag.

Mist or fog

Excessive spray mist and fog is an indication that either the paint used is too thin or the atomising air pressure is too high. It can also be caused by the improper use of the gun, such as incorrect stroking, or the gun being held too far from the surface.

Starving the spraygun

Starving a spraygun means literally what it says. Insufficient air or fluid reaches the gun which results in spitting, spluttering or a general fall-off in performance. Not enough air reaching the gun may be due to:

- Air transformer being clogged with rust or dirt.
- Air valves which are too small in size.
- Clogged air lines.
- Air hose or pipe line being too small in diameter.
- Inadequate air supply.

The gun may be starved of paint due to:

- Insufficient pressure on the tank.
- Too small a fluid hose.
- Too small a fluid tip.
- Fluid adjusting screw on the gun not opened wide enough.
- Paint being too heavy.

The application of metallics

A great deal of skill is required to achieve optimum performance in the application of metallics. Correct gun technique and evenness of the application will produce the very best results from a metallic finish. Contrary to popular misconception metallic finishes are no more durable or less durable due to the addition of the aluminium flake. Any colour can be turned into a metallic by adding silver base. There are many grades of silver and all the paint manufacturers offer a range of these base materials for use with their weight mixing systems to produce the many car colours on offer in the UK today.

When spraying metallics a 43 air cap is recommended to ensure an even and fine atomisation of the material. Also it will be necessary to overlap the gun stroke by up to 80% to ensure that 'banding' does not occur. Very accurate and even spraying is vital in order to obtain the best results.

Dark or wet patches of metallic finish can be 'mopped up' by dusting a dry coat in the particular area to soak up the excessive solvent. Care is needed when flatting a metallic finish for polishing as it is very easy to cut through the film and this will expose rings in the paint that cannot be polished away. Repainting is the only answer.

Application of basecoat and clear

This process is again highly skilful and requires great attention to detail.

The application of the base metallic needs all the care of a normal metallic but the application of the clear acrylic laquer can cause problems if intercoat times and application procedures are not adhered to.

The lacquer and the colour are two parts of a whole colour film and must be regarded as such. The lacquer must be applied within the operating window stated by the paint manufacturer. It is no good applying basecoat Friday afternoon and putting lacquer over the film on Monday morning. Intercoat adhesion can break down causing lacquer to flake away. Normally for TPA basecoat and clear, lacquer is sprayed within 15 minutes of the last basecoat colour. The film then dries out and contracts by solvent evaporation together causing a bonding through solvent penetration and migration.

The finish obtained by this process is very splendid and nearly all car manufacturers offer this system within their range of colours.

Repair of both metallics and basecoat is highly specalised and needs great care and attention to detail.

Repair of metallic finish

If you follow the procedure carefully then there is no reason why you cannot repair a damaged paint film to a degree where it cannot be detected.

Firstly, examine the vehicle in detail, noting the general state of the paint film, the age of the car, the general appearance of the car, and deciding whether or not it has been well looked after.

The area to be refinished is often dictated by the insurance assessor but unfortunately it has been my experience that paint matters are not exactly in the forefront of their knowledge and experience. Even if you are being pressed into painting one small area it is often better to make a decision to paint more of the car.

The procedure for a 'blow in' on an area, a passenger door for example, is as follows:

1. Repair the damaged area by panel beating or body filling.
2. Ensure that there is at least an area of bare steel between the body filler and the original paint film. Polyester fillers of any type applied over any paint film will lead to edging when recoated.
3. Ensure that the repair is perfect, as the owner will know exactly where he has sustained the damage.
4. Feather back the original paint film. Use a DA sander or wet flat with 240 or 320 paper.
5. Flat a larger area with 600 W/D.
6. Mask up the door and spray a primer coat after wiping the whole door with spirit wipe.
7. Spray filler coats over the area building up the bare metal and repair to the original finish level.
8. Allow to through dry.
9. Guide coat.
10. Use a block with 600 W/D paper to block out the repair and feather the primer edge into the original finish.
11. Compound the complete door panel except where the primer filler is present.
12. Check colour after thoroughly stirring the tin and emptying the contents into a mixing tin. Use the correct thinners to clean out the tin of any pigment trace left.
13. Thin colour to the correct viscosity using a BS cup.
14. Test colour on a 6 × 4 inch primed metal panel. When dry offer this up to the car for checking. In metallic finishes it is vital that the colour, appearance and texture are correct. If any one of these three is incorrect the owner will claim he can see a mismatch.

    Colour can be tinted, but only as a last resort when gun application has been altered to obtain the result.

    Appearance can be changed by gun speed, thinners, gun distance, temperature and general application ability.

    Texture can be altered in the same way.
15. When you are satisfied with CAT (colour-appearance-texture) then apply three single coats over the repair making each pass larger than the previous one, so enabling the solvent to break down the dry overspray on the job.
16. At this point the job can be left to through dry.
17. A 600 wet flat over the whole repaired area will set up an even substrate for a final colour coat.
18. Apply a single colour coat.
19. Using blending clear lacquer as 25% to the thinned colour in the gun apply a colour coat over the primary area.
20. Add 50% blending clear lacquer to the material in the gun and spray the secondary area.
21. Add 75% blending clear lacquer to the gun contents and spray the whole panel.

This achieves a fading out of the colour over the panel and by reduced opacity levels the human eye is unable to distinguish the overlapping that has taken place. The original finish on the panel shows through, so confusing the eye as to the boundary of the repair.

When dry, unmask the panel and polish with T cut, ensuring that you do not break through the film.

It is possible with care to make a repair in the metallic paint film which is totally invisible. This is the art of vehicle repair and should be practised at every opportunity.

Repair of basecoat and clear systems

This system of finishing has become more the standard than the exception and the repair is more difficult due to the addition of the lacquer coats which cause a development of the basecoat colour into a brilliance.

The procedure of finishing is the same as metallic but all the original lacquer has to be flatted off before repairs are carried out. This is due to the solvent-sensitive edges of the clear lacquer. Having made the repair good and flatted back base colour with 600 W/D, application of colour and then blending clear in the ratios of 25%, 50% and 75% is necessary. When this has set up then the clear lacquer must be applied within the time-scale indicated by the manufacturer. When the system has through dried it can be flatted with 1200 W/D and polished.

The areas for finishing and blending have been illustrated below. It is very important to use the design line of the vehicle to the advantage of the painter.

Often it makes good commercial sense to paint larger areas of the vehicle in order to offer an invisible repair to the customer. So often a paintshop will approach the job with blinkers and will not examine the wider aspects of the job in hand. Edge to edge finishing is not easy, and sometimes quite impractical, so the idea of blending large areas makes sense. It cannot be overstressed that the painter should look for natural breaklines or sharp changes in body curvature to finish or blend to.

With all of the modern refinishing materials offering a range of very high standard finishes, a painter can with confidence proceed into major colour coat work with the sure knowledge that the materials will hold up and last provided they have been applied as per the manufacturer's recommendations.

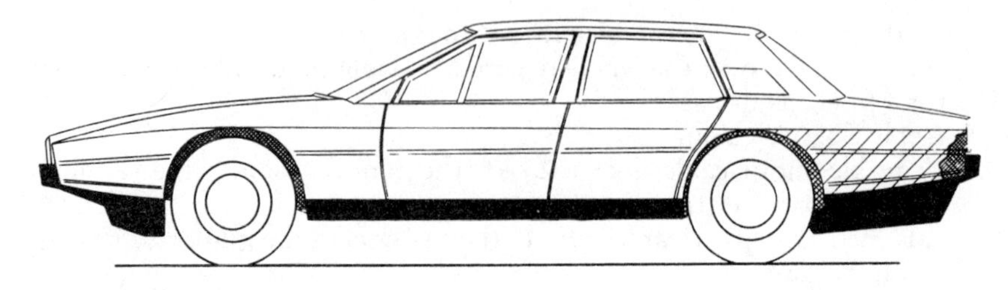

Fig. 19 Blending metallic to a wheel arch

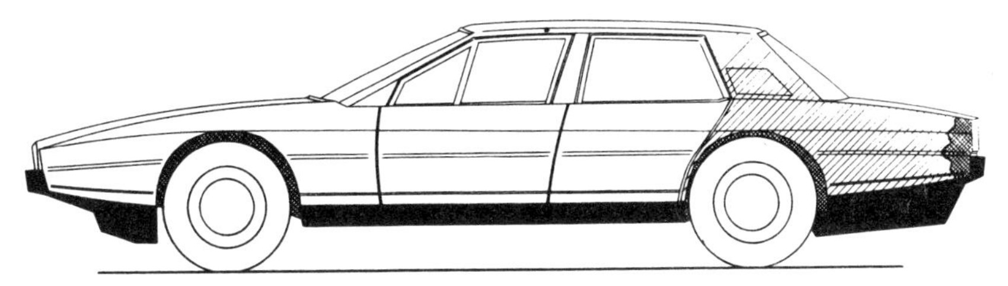

Fig. 20 Blending metallic to the C post

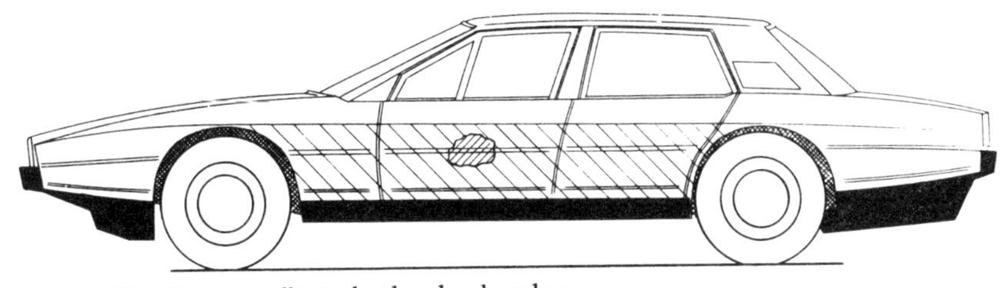

Fig. 21 Blending metallic to both wheel arches

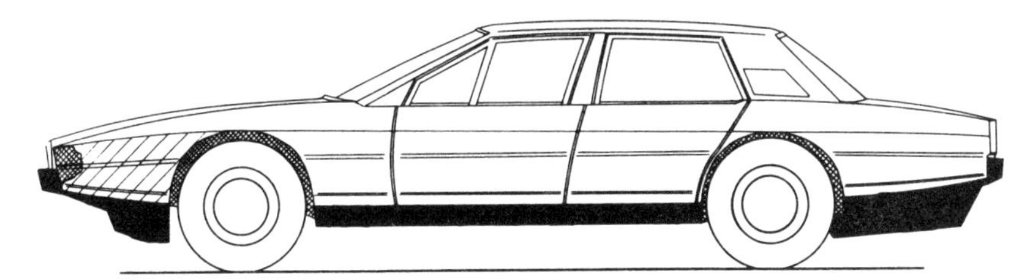

Fig. 22 Blending metallic to front wheel arch

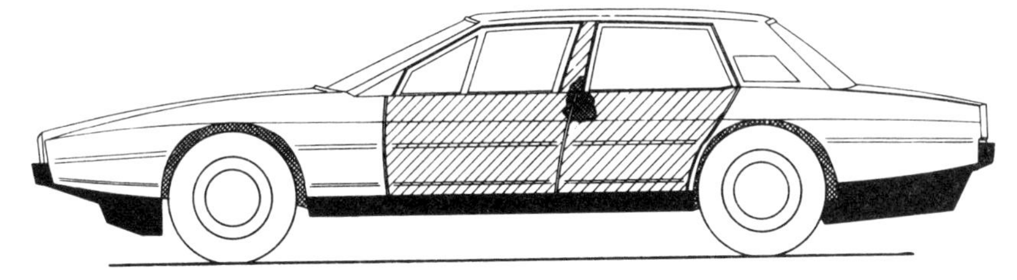

Fig. 23 Blending metallic to both doors

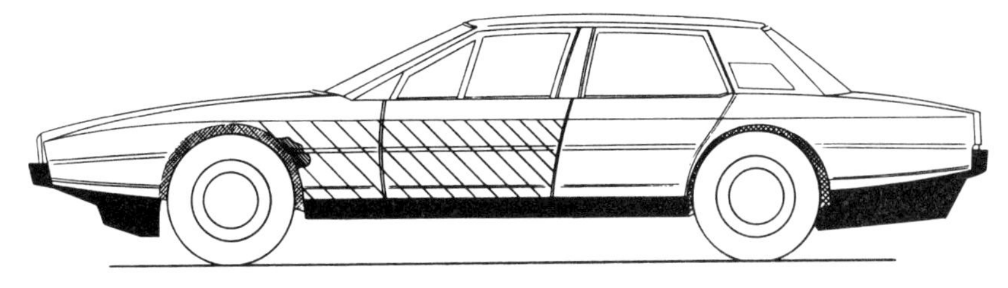

Fig. 24 Blending metallic to B post from front wheel arch

Examples

1. Damage on the rear quarter above the bumper but below the feature line. Blend out to the wheel arch. (Fig.19.)
2. Damage on the rear above the bumper and including the boot lid. Blend out whole quarter to C post and roof line. (Fig.20.)
3. Damage on the door panel below the feature line. Blend out to front and rear wheel arches. (Fig.21.)
4. Damage on the front wing. Blend out to the front wheel arch. (Fig.22.)
5. Damage to the B post and rear door. Blend out both doors. (Fig.23.)
6. Damage to the front wing behind the wheel arch. Blend out to B post. (Fig.24.)

Clear over base

The development of clear over base systems has brought about a further move towards harmonising the OE finish and the repair industry materials.

The polyester base paints are now closer in type and use than previous OE and repair materials. Because of these technological developments vehicle repairers are able to match more accurately the original finish, both colour wise and appearance.

Set out below is the full COB system of Ault Wiborg's Aultragem. This is a new high build finish that is comparable to the original factory finish.

Products

Aultragem basecoat – a range of 27 colour and tinting strength controlled bases.

Z4218	Stabiliser thinner – gives effective aluminium control to the basecoat.
Z4375	Two-pack acrylic clear lacquer for use over the basecoat.
Z6000	High solids activator for air dry or low bake/force dry.
Z4377	Standard thinner – for use with Z4375 lacquer in air dry or low bake/force dry applications.
Z4378	Fast thinner, air dry – for use with Z4375 lacquer in panel repairs and in cold body shops down to temperatures of 7°C.
Z4379	Slow thinner, low bake – for use with Z4375 lacquer in low bake applications, particularly where solvent boil or 'popping' may be a problem.
Z4380	Fade out thinner – for use in panel and local repairs.
Z4381	Single pack acrylic clear lacquer – for use in panel and local repairs.
Z4382	Tie coat – an adhesion promotor for Z4381 lacquer.
ACA 100	Anti-crater additive – a silicone-free additive which can reduce the risk of cratering (cissing) re-occurring.

Substrates

Original equipment, Aultrafix and Aultrabuild primers and all finishes in good condition. All surfaces must be well flatted and free from contaminants, prior to commencing spraying. Aultragem can also be used with the Aultraflex system in the painting of plastic substrates.

Aultragem basecoat

Mixing ratio by volume: 1 part Aultragem basecoat + 1 part Z4218 stabiliser thinner.
Viscosity @ 20°C: 20–23 secs BSB4, 14–17 secs DIN 4.
Spray pressure at gun: 60–65 psi DeVilbiss JGA 43 or 80 equivalent. A 10 m air hose, even in good condition, can reduce air pressure by as much as 10 psi.
Application: Apply 2 or 3 coats, with a 5 minute flash off between coats. Certain colours may require additional coats to achieve full opacity.
Drying @ 20°C: Allow final coat to flash off for 20 minutes, before applying activated and thinned lacquer (Z4375). The basecoat can be masked up after 20 minutes for duo-toning work. Basecoat can be denibbed after 20 minutes if required, to remove minor imperfections. Aultragem basecoat must be overcoated within 1 hour of application.
Usable pot life @ 20°C: Indefinite. Stir and recheck colour before re-use.

Two-pack acrylic clear lacquer (Z3475) air dry

Mixing ratio by volume:
Standard air dry: 4 × Z4375 lacquer + 1 × Z6000 activator + 2 × Z4377 standard thinner.
Fast air dry: ideal for use in cold body shops down to a temperature of 7°C. 4 × Z4375 lacquer + 1 × Z6000 activator + 1 × Z6000 activator + 2 × Z4378 fast thinner.
Viscosity @ 20°C: 17–19 secs BSB4, 14–15 secs DIN 4.
Spray pressure at gun: 60–65 psi DeVilbiss JGA 43 or 80 or equivalent. A 10 m air hose, even in good condition, can reduce air pressure by as much as 10 psi.
Application: Apply 2 or 3 coats, allowing 5 to 10 minutes flash off between coats.
Drying @ 20°C: For a 50 micron film. Drying times will be significantly influenced by shop temperatures and film build.
Standard air dry: Dust free – 15 to 20 minutes. Hard dry – 16 hours. Recoatable – 16 hours.
Fast air dry: Dust free – 15 to 20 minutes. Hard dry – 2 hours. Lacquer can be flatted to remove dirt particles, etc., and overcoated with lacquer.

9

Undercoat Systems

Adhesion must be one of the most important factors in the life of any paint film. There is the adhesion of the first coat of self-etch primer to the steel, aluminium, or GRP; without this sound base to work on paint failure can occur at any time. Good adhesion is necessary in the intercoat stages, i.e. filler to self-etch, and finally the adhesion between a sealer coat and colour. All these applied films of material bond into one complete system to protect and give a lasting decorative effect to the finished vehicle.

The procedure to build up a good substrate is as follows:

1. Abrade the surface with 180 wet & dry.
2. Wash down and dry.
3. Apply metal treatment (Deoxidene) which is a phosphoric acid solution.
4. Wash down with hot water and blow dry.
5. Apply 1 single coat of self-etch primer.
6. Spray three coats of filler, either nitrocellulose-based or two-pack.
7. Allow to dry or low bake (80°C panel temperature for 15–30 mins).
8. Guide coat and flat back with 600 wet & dry.
9. After careful checking of the surface, stop up with two-pack stopper.
10. Block down stopped areas.
11. Apply sealer coat or further coat of filler.
12. Apply colour coats for first paint.

High build materials of today are extremely good and it is possible to build a film of 8 thou (approx. 200 microns) very quickly. Care should be taken not to exceed 8 to 10 thou in total film build.

Contrary to popular belief the thicker the film the more chance there is of breakdown and cracking. Motor manufacturers of today end up with a total film build of between 3 and 5 thou (75 to 125 microns approx.). This thinner film allows the body to move and distort with road shock without damaging the paint film. The elasticity present gives the necessary movement.

The systems

An undercoat system is a term which describes all coatings which provide the substrate base for the finishing colour coats. Adhesion and build are the main features of these materials.

Primers

Primers are designed to give the first adhesion to the steel, aluminium or GRP panels of the car. Ideally a self-etch primer should be used. This is zinc chromate with orthophosphoric acid and when it is mixed and sprayed onto a panel the acid penetrates the outer surface of the metal and carries the chromate with it, so ensuring a key for subsequent filler coats. Primers are effective as thin, lightly applied coats, and they are not designed to be flatted. All primers are formulated to give maximum protection from corrosion, and they range in type from nitrocellulose to synthetic base.

Primer surfacers

Surfacers have to fulfill two functions: firstly to provide good adhesion to primers or original finishes, and secondly to fill scratches. When the filler is dried through then flatting can take place to resurface the film. By flatting out imperfections a sound and smooth base is ready to receive the colour coats.

To achieve high build and better flatting large amounts of pigment are used, which always leads to a massive intake of water into the film during flatting operations, and afterwards in service. It is important to thoroughly dry out primer and filler coats before colour application and to ensure that at least 2 thou of film build is left after flatting.

Fillers

These materials are heavily pigmented to fill deep scratches and panel irregularities to a greater degree than primer surfacers. The two-pack polyester fillers are now very popular as they can be applied wet on wet and will, because of the catalyst action, give a huge film build with virtually no sinkage. Force drying or low baking can, depending on temperature, give a fully cured film within 15 min. @ 80°C. Longer through dry times are directly related to the heat available at the panel. These fillers will flat back extremely well and will take any sealer or colour system available.

The importance not only of a good primer filler substrate to work on, but also of the right primer filler base, cannot be emphasised too strongly. Plan the job and think carefully about the finish that is being used and the vehicle it will be coating. So often the wrong materials are used somewhere in the colour finish system and the results can be difficult to rectify as well as displeasing to the customer. It is important to keep to the manufacturer's specification and follow it carefully. All paint manufacturers spend a great deal of time and money ensuring that their system works well, is easy to operate and stands the test of time. Pay them and their paint technicians the compliment of following their carefully prepared instructions. Success with every job is the reward for following this course of action.

10
Finish

It is the finish of a motor vehicle that is actually observed. The smoothness, gloss reflection and colour will give instant visual stimulation and usually awake a desire to possess.

The process of achieving the ultimate finish can be divided into two stages. Firstly, the application of the finish colour and lacquer, where used, and secondly, the work carried out after the film has cured, to achieve the maximum gloss level and smoothness.

The standard finish of all the major paint manufacturers is excellent. The gun finish that can be obtained by proper and careful application is more than satisfactory for all mass-produced motor vehicles. However, should you wish to proceed beyond that level, a great deal of hard work and time must be expanded to achieve the ultimate.

Some finishes do not lend themselves to further work and are primarily designed to give excellent gloss and flow out from the gun. These, generally, are the splendid high build two-pack acrylic polyester materials presently available on the market. The Aultragem COB polyester system manufactured by Ault Wiborg Paints Ltd is an example of this advanced technology. The product was developed to meet the increasing use of clear over base finishes by motor manufacturers. It is designed to finish out with brightness and brilliance straight from the gun. The two-pack lacquer has a high density hardener, gives high build, excellent durability and super-quick drying.

The material used by Aston Martin Lagonda in production is a TPA (thermoplastic acrylic) made by Ault Wiborg exclusively in Aston Martin's own colours. This material is extremely durable, light fast and very polishable. The excellence of the material is matched by the skill of the paintshop in working up the cured film to a finish which is the finest available on any motor vehicle produced in the world.

11

Metallic Finish

The rise in the popularity of metallic finish since the early '60s is nothing short of phenomenal. Market forces and overall demand acting on the motor manufacturers have forced them to produce more and more metallic colours. Contrary to popular belief it is the styling department at the manufacturers that designs the colour and the paint manufacturers have to follow and produce the colour for line use and refinishing. Currently there are some 18,000 car colours on offer spanning the last ten years. Metallics and basecoat and clear are on the increase simply because well finished and attractive cars sell well.

Metallic colours are a combination, in varying proportions, of straight colour and aluminium flakes of various sizes. Simply put, the more metallic content the lighter and more silvery the appearance; the coarser the grade of silver the brighter the colour appearance. When repairing metallic finishes the colour, appearance and texture must all match the original, otherwise the customer will be able to see the repair.

In earlier times only one grade of aluminium was used and this made matching a little easier, but today many colours have combinations of several grades of aluminium which makes the process a little more complex. Spray technique and careful attention to solvents, viscosity and air pressure will make the difference between success and failure in metallic matching.

Light reflection

When daylight falls into a metallic finish, the aluminium particles act like small mirrors and reflect the light out. Because of this reflection, the maximum amount of reflected light will be observed when the finish is viewed at right angles. As the viewing angle decreases, so will the reflected light. This means that viewed from the side the colour will appear darker than when viewed at the face. These face tones and side tones play an important role in the matching of metallics.

The differences between face and side tones will be most apparent when the aluminium flakes are in parallel to the surface of the paint film. If the flakes of aluminium are more randomly spread through the film build the face tone will be darker and there will be less difference between the face tone and side tones. To accomplish an acceptable colour match the latter method is the preferable route. Differences between the face tone and the side tone are hard to rectify.

Colour matching

This is the most difficult area of refinishing. There are so many permutations of cause and effect that can mislead the most experienced and skilled painter. However, following good basic procedures and thinking and planning the job through will in most cases bring the job to a satisfactory conclusion.

Firstly, when matching an original finish with a nitrocellulose or acrylic finish, it must be remembered that the build and opacity of these materials is far less than factory finish. Original finish will have up to 50% solids, whilst repair material will only have 28% to 32%. This means that the refinish material solids tend to float in the film, giving a different appearance.

Spraying methods which produce thin coats that dry quickly will allow the aluminium flakes to lie parallel to the paint film surface which gives the greatest contrast between face and side tones. Thick, slow drying coats of colour allow the aluminium flakes more movement and therefore a more random orientation in the paint film – which gives less difference between the face and side tones. Better control over face and side tones can be obtained with nitrocellulose and acrylic lacquers than with the slow drying synthetic finishes, simply because of the wider range of application viscosities at which lacquers can be sprayed whilst still controlling the aluminium appearance. The difficulty of edge to edge repairs in metallic makes the use of the fade out and blending lacquer techniques highly desirable. Having matched a metallic colour as closely as possible, blending will make the repair impossible to detect. This ensures total customer acceptance.

Tinting metallic colours should be carried out as a last resort. The paint manufacturers will have ensured that the colour is totally correct by use of a computer to the motor manufacturers master panel. This does not necessarily mean the colour will match the car, as many production problems within the car plant can cause a difference. When the decision to tint has been taken, then remember that both the face tone and side tone will be affected.

It is important to follow the guidance of the paint manufacturer when it comes to tinting. A whole table of effects is available from every manufacturer. For simple guidance, however, aluminium affects the face tone, white affects the side tone and permanent blue, green, yellow, red, maroon and black effect face and side tones.

Always tint with reduced strength tinters and slowly allow time for colour development after spraying. Always use at least a 6 × 4 inch panel for testing, so you can see the results. Follow the paint manufacturer's instructions with metallics and pay great attention to conditions and techniques. Think and plan the job through. If you hit a problem then stop and think, even leave the job. Do not hurry on, piling on more material, hoping to bury your problem. Remember, only doctors can do that.

Pearl effect finishes

A further new development that has taken place recently is the introduction of pearl effect finishes. These are very attractive and eye-catching and certainly they can enhance a vehicle to a stage where it totally transforms the car. The Tickford Lagonda shown in Fig. 25 proves the point admirably. This car has been painted in the Glasurit material, Glassomax, and this is now used in production at Aston Martin Lagonda when pearl finish is ordered by a customer. The process consists of a white primer surfacer fixed with a hardener, followed, after wet flatting, by the pearl effect colour, and then the final application is with acrylic lacquer, again fixed with a catalyst hardener. The end result is a very hard, durable and attractive finish. I am sure the popularity of this type of finish will grow. At present there are five pearl effect colours. They are all base white but the side tone can be red, blue, yellow, green or white. They are particularly effective on rounded panels which allow the shading of the side tone into the face tone, which gives a stunning appearance to the car.

Fig. 25 Pearl finish on a Tickford Lagonda

These finishes can be repaired but care and attention to detail are the hallmarks of a successful job – a case in point where the very full and thorough instructions of Glasurit Valentine need to be followed carefully and to the letter.

In my experience all of the paint manufacturers offer excellent literature and thorough instructions in the use of their materials. To add to that, all of them offer excellent demonstration and technical service teams who will come and investigate a problem or demonstrate the products on your premises. All of them run training courses for paintshop staff to attend, and these are both interesting and informative. No one is more aware than the paint manufacturers that the repair and refinishing industry needs help and guidance when attempting to use the new technology products that are part of the modern motor vehicle.

12

Rectifications

Faults that occur in a paint film are almost always the fault of the operator. This is a fact that vehicle sprayers and finishers with little experience so often find hard to accept. Painting materials and systems are rarely incorrect or below standard. The reason for this is simply that all the manufacturers test material in production, random samples are taken also at the pre-filling stages, and then batch samples are taken and retained after filling. Every manufacturer can give you all the details of a production batch of material and prove the testing that has taken place for colour, opacity, gloss and film build. With such care in the production of paint and the elaborate exposure testing of paint films to ensure durability and long-lasting weatherability it is almost impossible to receive faulty paint.

It is obvious that paint faults lie in the conditions and application of the materials. Let us examine the faults, and their causes, that are most common and then the rectification of these faults.

Fault: Air blowing. Low stoving problem that occurs when air is trapped under a paint film in stopper, body filler or in the gel coat of GRP, and expands up through the film.
Cause: Failure to apply stopper or filler correctly. Poor application of gel coat.
Rectification: Strip to bare metal or GRP (use mechanical method) and fill correctly.

Fault: Blooming.
Cause: Spraying in cold or humid conditions with poor air movement. Lack of proper mating.
Rectification: Spray further material or polish when dry.

Fault: Silicone or grease contamination.
Cause: Use of silicone polish on vehicle. Oil or grease from compressor or soap used in flatting operations.
Rectification: Strip to bare metal.

Fault: Cracking or crazing.
Cause: Excessive build in paint film.
Rectification: Strip to bare metal.

Fault: Dirt inclusions.
Cause: Dirty or non-existent filters in spray booth. Contamination by body fillers, or road dirt of vehicle. Dirty or dusty booth.
Rectification: Flat and polish or repaint area.

Fault: Dry overspray.
Cause: Use of incorrect thinner. Gun too far from panel. Air pressure too high. Incorrect gun set-up.

Rectification: Allow to dry and wet flat, check for porosity.

Fault: Flotation.
Cause: Heavy application of wet film, causing the metallic particles to float in the film and show as dark lines.
Rectification: Either allow to harden and flat or in mild cases dust the surface with colour to allow material to soak up dry spray.

Fault: Poor gloss.
Cause: Use of incorrect thinner. Sinkage in primer filler. Poor spray extraction. Poor drying conditions, cold and damp.
Rectification: Allow to through dry and 1200 flat and polish.

Fault: Orange peel.
Cause: Excessive paint application, incorrect thinner. Poor application technique. Low air pressure. Incorrect viscosity.
Rectification: After through dry, flat and polish or respray.

Fault: Adhesion break down.
Cause: Self-etch primer not used. No metal pretreatment used. Use of incorrect thinner. Poor flatting – no key. Contamination by silicone, wax, oil, grease or water, soap or welding flux.
Rectification: Strip to bare metal and respray.

Fault: Runs
Cause: Poor spraygun technique, air pressure too low. Application speed too slow. Over-thinning material.
Rectification: Wet flat when dry and polish.

Fault: Sinkage.
Cause: Excessive paint application. Insufficient drying time between coats of primer filler. Dry spray of primer filler. Poor repairs not feathered out enough.
Rectification: Allow to dry through and flat and polish or respray.

Fault: Striping in metallic.
Cause: Failure to overlap by at least 60%. Incorrect gun setting. Spray too narrow. Failure to hold gun at right angles.
Rectification: Either allow to through dry and lightly denib and respray correctly, or dust light coats of colour onto the panel and allow the dry spray to act like blotting paper and soak up the solvent in the dark stripes. Then allow paint to flash off and respray correctly.

13

Final Finish

The final finish of a vehicle is all that is seen by a customer. The preparation, the priming and filling are unimportant to him provided the final finish is of full gloss and free of imperfection. Today, society is ever seeking and expecting perfection. In the early days if a person owned a car, it was a wonder, and if it actually went down the road and all four wheels went round together, he was satisfied. If you stop for one moment to look at the technology that we take for granted now, the acceleration of expectancy that will catapult mankind into the 21st century is frightening. I hope the technicians behind the scenes will cope, but if they cannot then I am sure the marketing men will be able to hold the line until they do.

The finish and the trim are the items of prime importance, simply because they are seen by the owner and his wife. The application of the finish coats of colour is of great importance and the techniques employed by the operator will at this stage determine the final outcome.

Full respray 'A'

Straight colour application in nitrocellulose, acrylic or two-pack polyurethane.

1. Well flat vehicle with 600/800 wet and dry.
2. Wash vehicle down thoroughly with water.
3. Dry vehicle thoroughly. Ensure all gutters, bonnet, boot and door returns are blown out and wiped.
4. Remask vehicle.
5. Wipe vehicle down with intercoat spirit wipe.
6. 'Tak off' using good quality tak rags.
7. Blow out vehicle finally.
8. Ensure good air movement and booth temperature (20°C).
9. Mix paint to manufacturer's spec., using the correct thinners and viscosity cup.
10. Strain the paint when mixed.
11. Fill the gun pot and test fan spread and atomisation.
12. Check air pressure at the gun.
13. Apply single coat for 'set up' over whole vehicle.
14. Spray the roof, down to one wing. Work round the vehicle methodically. Spray boot and bonnet separately.
15. Overlap gun strokes by 50%.
16. Ensure accurate trigger off to avoid unnecessary overspray panel to panel.
17. Check gun distance from the panel all the time.
18. Work to rhythm of speed and distance.
19. Adjust speed of traverse to flow of material.

20. Attempt to encapsulate the vehicle in material to a uniform depth.
21. Apply second full coat of material after advised 'flash off' time. Repeat the application with all the care previously outlined.
22. At this stage a wet on wet coat can be applied. This is where a full coat of colour is sprayed and then another full coat is applied with no flash off time. This allows a very heavy film build which 'flows out' to give a superb finish. The drawback is that a run or sag can easily occur.
23. Allow the colour to flash off and settle before applying heat. Follow the manufacturer's recommendations on this.
24. Unmask immediately after low baking or force drying. Watch for any material that may have bridged onto the tape. Cut this with a blade if it has occurred.

Full respray 'B'

Metallic colour application in nitrocellulose, or acrylic.

As in 'A' 1 to 10, then:

11. Use a 43 air cap or equivalent on the gun. Test for fan spread and atomisation.
12. Ensure that the fan is 'full' and that there is no area of lighter film weight. It is very important to avoid this, as it will transfer to the job a 'tramline' effect.
13. Spray the roof and down to a panel and work methodically round the car.
14. Overlap the gun strokes by up to 85%. Metallic colours vary, for instance some light blue colours will tend to 'band' or 'float' very easily and they need a good overlap to avoid this. Proceed as 'A'.
15. If on the final wet on wet coat 'flotation' or 'banding' does occur, then make a further application of material traversing the panels at a faster speed and further away. This has the effect of 'dusting' the finish with drier material which tends to soak up the solvent laying in the film like ink into blotting paper. The finish will settle and flow out giving the uniformity which is so important with a metallic finish.

Full respray 'C'

Basecoat and clear application in nitrocellulose or acrylic.

This operation is the most demanding because of the extra care and attention to detail needed at every stage. The application of the metallic base is quite straightforward but it is most important to watch for any dirt inclusions or contamination in the last coats. To spray the clear lacquer over the base with inclusions traps them and there is no way to rectify this without flatting back when the film is through dry. Any dirt must be eradicated before lacquer coats and if necessary the metallic base resprayed.

It is most important to follow the manufacturer's instructions with regard to intercoat flash off times. Normally clear lacquer is applied approximately 10 to 15 minutes after the last metallic basecoat has been applied.

Applying lacquer can be hazardous as the general rule is to go for a high film build as quickly as possible. Double header coats are usually recommended and this does put the risk of runs and sags to the forefront. Proceed with care and all will be well. Do not be afraid to spray lacquer coats more lightly or, if too much has been applied too quickly, leave much longer between coats to allow solvents to flash and the film to settle down. One can always apply more material; it is a lot harder to take it off!

Two-pack polyester acrylic finishes are very much the new materials and spraying this finish is different due to the advanced technology that enables the manufacturer to build in high solid content for maximum opacity and film build. These paint finishes do spray in a slightly different way and generally they are easier to use. All the principles outlined earlier still apply but the actual spraying is better due to the improving technology that surrounds and supports these materials. Higher opacity levels and film builds, coupled with good 'flow out' properties overcome many of the problems that surround nitrocellulose and acrylic finishes.

Fine finishing

It is almost impossible to finish a complete vehicle and not have at least one blemish, dirt inclusion or small run present in the final film. It is always better to leave the blemish or dirt inclusion if it cannot be polished out rather than flat and refinish the area. Nitrocellulose materials as well as acrylic finishes are not inert and will break down in their own solvent, which simply means that the surface area around the imperfection will soften when colour and solvent are re-applied. This has an effect on the long-term life of the film, and if this action can be avoided it should. A great deal of time and effort can be expended quite unnecessarily, and it is often better to leave well alone.

Final finish is really the little extra effort that is put into the operation to give the most complete paint film; to achieve a very high gloss level and a total reflective appearance. Only Aston Martin Lagonda and Rolls-Royce strive to obtain this and consistently succeed. No other motor manufacturer even wishes to attempt to reach these standards. This is due to the numbers of vehicles being produced and the intensive labour requirement to accomplish it.

The key to achieving the ultimate motor car finish can be quickly summed up as great skill and great care at every stage of the painting operation, great attention to detail, a great many hours of dedicated work and great pride in the job.

14
Contamination

Modern car finishes are very resistant to all normal forms of atmospheric attack. Provided a simple maintenance procedure is followed they will retain their gloss colour and protective properties throughout the life of the vehicle. However, car finishes are not chemically resistant. Severe local contamination of an acid or alkaline character can occur. If it is left in contact with the paint film for any length of time it may cause pitting and colour change.

Sources of contamination

Contamination may come from three sources:

- Airborne industrial fall-out and acid rain.
- The vehicle itself.
- The highway and adjacent ground.

Industrial fall-out

The term industrial fall-out was first used to describe minute particles discharged from the chimneys and workshops of the iron and steel producing and fabricating industries. Fall-out may be heavy near manufacturing industries but it is in no way confined to such areas. Strong winds can carry the effluent many miles. Railway networks are another source, the braking of rolling stock producing a discharge of iron particles.

In the presence of moisture, and particularly if they are in the least bit magnetic, the particles will become attached to the paint surface. Soon they may discolour (rust-coloured spots) and pit the surface of the paint. If allowed to remain they may penetrate right down to the primer or even the metal substrate itself. Regular and requent washing is the best safeguard against attack, but if attack does occur:

- Light contamination, when the particles are not embedded in the film, may be removed by compounding and polishing.
- Heavy contamination can be removed by a chemical wash using a 10 per cent (54g) solution of oxalic acid to ½ litre of water. Apply the solution taking care to prevent it from running behind mouldings, etc. It is important to keep the surface wet and active by several applications of the solution over a period of 15 to 20 minutes. Thoroughly wash off all traces of the solution and dry off. Then polish as necessary.

Acid rain

Acid rain is the term given to rain containing effluents from manufacturing chemical industries, particularly power stations. Some of the effluents may be

Sources of contamination	
Industry	Chemicals used
Fertiliser	Sulpuric acid nitric acid Phosphoric acid superphosphoric acid nitrogen compounds
Cement	potash–potassium hydroxide lime-calcium hydroxide
Paper mill	sulphuric acid lime
Distillation of hardwood	acetic acid
textile fibres	caustic soda other caustic compounds nitrogen compounds
Petroleum	sulphuric acid
Chemical	almost any possibilities including solvents
Dyes	sulphuric acid hydrochloric acid nitrogen compounds hydroxides nitric acids
Pigments, paints, varnishes	sulphuric acids
Lacquer, printing inks	caustic–sodium hydroxide
Soap and detergents	caustic soda potassium hydroxide caustic-sodium hydroxide
Steel, copper, lead, zinc	caustic-sodium hydroxide
Tin and mercury	sulphuric acid

Effect of chemicals on certain colours		
Colour	Appearance	Chemical cause
Yellow	White spot	hydrochloric acid – muratic acid
	dark brown spot	nitric acid
	red spot with film degradation	sodium hydroxide–caustic
	spotty blistering	acetic acid
Non-metallic medium depth blue	slight whitening	nitric acid
	slight whitening with film degradation	sodium hydroxide–caustic
	spotty blistering	acetic acid
White	pink	nictic acid
	pink colouration with film degradation	sodium hydroxide–caustic
	spotty blistering	acetic acid
	yellowing (acrylic lacquer showed no discolouration)	ammonium hydroxide
Medium depth blue	slight light blue spot	hydrochloric acid
	dark blue spot	nitric acid
	deep purple spot with film degradation	sodium hydroxide–caustic
	spotty blistering	acetic acid

Fig. 26 Sources of chemical contamination

acidic or alkaline in the presence of water, e.g. sulphur dioxide will dissolve in water to give an acidic solution, whilst a mixture of cement dust and water is strongly alkaline. Such effluents will attack paint films. The attack may take the form of discoloured spots due to attack on the pigment (e.g. some reds will develop a blue tone if attacked by acids and a brown discoloration if attacked by alkali) or distortion of the paint vehicle itself.

Some pigments are more prone to attack than others. Brunswick greens are well known for their proneness to attack by alkali (yellow discoloration). The aluminium flake in metallic paints is particularly prone to attack by both acid and alkali. Basecoat and clear metalic finish has the advantage that the clear coat shields the aluminium from the contaminant. But even clear paints may be attacked, losing transparency and/or gloss. Air dry paints, particularly when new, are more vulnerable than stoved finishes, but become more resistant to attack as they age. Of the present range of refinish topcoats, best resistance to acid rain attack is shown by the polyurethane paints.

Contamination also comes from agricultural and horticultural sprays. Certain insecticides such as DDT, dieldrin and malathion can cause spotting and pitting or blistering of paint in the presence of moisture and warm sunshine. Some concentrated herbicides are literally paint removers. Bird droppings can cause severe distortion and/or discoloration of paint films.

In the United States, the residues of some types of dead fly, when baked on by the sun, have caused discoloration and cracking.

The vehicle itself

Contaminants from the car include petrol, oil, grease, brake and de-icing fluids, which may result in staining and/or breakdown of the film. All these contaminants may be present on the highway itself, as may exhaust wastes, tar and other road building materials and residues from various spraying operations, e.g., creosoting of fences, cropspraying.

Tar, bitumen and grease stains can be removed by wiping with white spirit or similar suitable solvent, but light contamination from acid rain needs compounding and polishing. Heavier contamination may require that the vehicle be painted – wash thoroughly, clean with a solvent-based water miscible cleaner, wet flat, dry off, spirit wipe and repaint. In extreme cases it may be necessary to strip to metal and repaint.

Regular and frequent washing is important to prevent chemical attack, particularly when the car is new or has been freshly painted. Some degree of protection may be obtained by regular cleaning and waxing. Better protection will be provided by non-silicone polishes, but there is no guarantee of freedom of failure through contamination.

Figure 26 sets out a table of some of the sources of contamination and the effect they have on a paint film.

15
Care of Film

This is a most important aspect of a fine finished vehicle. The after-care of a finish will give the lasting service that an owner should demand.

Always wash the vehicle in warm water with the addition of a liquid soap, about a teaspoon to a gallon or 5-litre bucket. This is mild enough as a wetting agent to remove road film without damaging the paint film itself. The vehicle should always be wetted up with warm water before attempting to sponge it down. A great deal of damage is done by people washing mud and grit off with a sponge on a dry surface and it is almost the same as wet flatting the vehicle.

Thoroughly clean the car with several applications, especially in winter months when the salt and grime is so corrosive. Wash your car down as often as possible in this way. After this operation hose the car with a low pressure cold water hose to remove all traces of the liquid soap. Hose under the car as much as possible to break down the salt.

Leather the vehicle dry with a good quality chamois, taking care to rinse out often in warm, clean water.

A paint film needs waxing two or three times a year with a good quality, non-silicone wax. Before this operation it is important to wash and then to clean off the road film by using T cut. It is most important that this is done as it is unlikely that washing in warm soapy water will dispense the road film completely. After T cut, apply two coats of wax and polish to a high lustre.

Wash your vehicle often, it improves the film. As you will observe,

Fig. 27 An Aston Martin Volante awaiting shipment to the USA, coated in ICI Tempro protective coating. This product, which is easy both to apply and remove, has proved invaluable in protecting the fine finish from shipborne contamination. Every Aston Martin and Lagonda is coated with Tempro after final painting and examination

chauffeur-driven limousines are always immaculate due to continuous washing and leathering.

Many of the motor manufacturers are now protecting their paint finishes with a new acrylic coating called Tempro, developed by ICI. This material is applied in one coat and will keep almost any contaminate out of the paint film during the delivery or storage time. Tempro is removed with a special alkali-based material that breaks down the coating without damaging the paint film underneath. It is then washed off to reveal the original finish. Aston Martin Lagonda Ltd use this coating on every vehicle to protect the fine finish during delivery time to the point of sale anywhere in the world (see Fig.27). It has been particularly good when vehicles have been sent by sea to the USA or Middle East.

Tempro is another example of how paint technology is advancing to give the highest possible standard of protection to a paint finish and the vehicle to which it is applied.

16

Future Finishes

For some time now there has been a general thought that the paint industry would develop into one area for both OE and refinishing.

It would appear that the ideal paint process would consist of a two-pack primer filler with high etch qualities, followed by a two-pack high build acrylic urethane colour film. The total film build would then be either high baked (320°F) or low baked (176°F). There are enormous advantages to using a simple system such as this but possibly paint technology will need to develop further and the end user will have to be satisfied with a finish that might be judged as only just acceptable.

However, as the world develops and, hopefully, the man in the street has more disposable income, the consumer sales idea of simply changing your car every year or so will become more prevalent and a lower standard of finish will be acceptable provided that drive, economy and total reliability improve even further.

It is a fact that many nations have a love affair with the motor car and I believe many British males would take the thing to bed if they could get it there. Our continental cousins are more objective towards their mode of transport and tend to regard it as a tool of modern life to transport them from one place to another for work or pleasure.

Generally the finish on European vehicles is good, robust and very serviceable but it is not a high controlling factor in the purchase requirement. I therefore believe that as the two-pack urethanes become more widely used and acceptable, the British will follow the continent and generally the motor vehicle finish will be slightly down-graded over a period of time, with economy and reliability becoming the factors of most concern in Britain.

The finishing and refinishing of modern production vehicles has become and will continue to become easier and less skilled as time progresses.

I am sure that the pre-colour detachable panel vehicle will be a realistic possibility in the future and this will revolutionise the cosmetic panel repair business in Europe. Body shops will still be in business repairing the structure of the cage that carries the mechanical parts and the outer panels. Colour control of the impregnation will have to be very accurate but the bonus will be a vehicle that is easy to repair and will keep up its structural strength even after several years on the road.

Developments in carbon fibre and all of the range of polycarbonates make a totally 'plastic' car a very real possibility in the future. We can be assured, however, that as long as people drive cars they are bound to bump into each other and the repair shops around the world can sleep safely with their futures certain.

17

Management of Paintshops

The skills and judgements necessary for the successful management of any production unit have been well listed, outlined and developed by many authoritative writers. However, it has been my experience in the USA, Great Britain and Europe that paintshop managers or middle management responsible for the paintshop have shown repeatedly by their actions, theories and questions to me, that their understanding is a grey area indeed, in fact a dark grey area.

The motor industry has persistently appointed motor engineers or motor body men to management positions overseeing the paintshop operation. With the best will in the world, these disciplines do not readily lend themselves to the smooth and effective operation of a paintshop.

Effective management in a paintshop consists not only of programming the workload and the staff hours, but of being the leader who can answer the technical problems as they arise on the shop floor. He must be able and free to make commercially effective decisions on behalf of the company. For example, it is better to strip a vehicle to bare metal and repaint in the sure and certain knowledge that the job will last for many years, rather than to take a decision to spray a sealer coat over the problem area, refinish, get it out to the customer and hope it does not come back.

I know that there are a lot of 'expedient' paintshop managers with 'fingers crossed' situations, and overall the warranty claims lists of dissatisfied customers and 'litigation pending' prove their lack of professionalism, naïvety or both. They do the motor industry no good at all and in fact strengthen the public's preconceived notions that all garages and refinishers are untrustworthy and unreliable.

This is an area in to which senior executives must look carefully and either train management with engineering backgrounds to fully understand paint and the associated problems and future developments, or better still, train first class technically proficient painters to become managers.

A sensible, reactive management team, selected for their knowledge and ability to run the mechanical service and repair, the body shop and the paintshop, and responsible to a works director, is the certain proven path to successful and profitable operations.

Communications and joint decisionmaking is critical to overall success. There has been, and I suspect always will be, a thin dividing line between where the body shop finishes and the paintshop begins. Good communication and understanding between the two managers or foremen will alleviate the special 'them and us' syndrome which appears wherever you go. (This, I might add, is not solely confined to Great Britain.) Once sensible lines of demarcation are clearly read and understood by both sides, then work flows successfully between shops without the dreaded 'back working' of vehicles.

A manager who is economically successful must:

- Be able to manage by objectives.
- Be prepared to take more risks, and his decisions will have to be made at all levels in the organisation.
- Be able to calculate the risk, choose risk alternatives, and establish in advance what is likely to occur.
- Be able to make strategic decisions and both present them and follow through.
- Be able to build an integrated team, each member being capable of managing and imposing his own supervision for quality, quantity and effectiveness.
- Be able to communicate quickly and effectively, to inform and motivate people.
- See the business as a whole. It is no use managing in isolation. The overall good, profitability and advancement of the company must be uppermost in his considerations.
- Manage with integrity. The respect from his staff, other managers and executives contributes entirely to his success and the ultimate success of the company.

As the finishing and refinishing of motor vehicles heads in time and technology towards the 21st century, so the management teams now and in the future must realise the importance of the right man in the team, technically qualified, enthusiastic and with the integrity to make it happen. Financial success is the reward if you get it right.

Index